NOT GOOD ENOUGH FOR CANADA

Canadian Public Discourse around Issues of Inadmissibility for Potential Immigrants with Diseases and/or Disabilities, 1902–2002

Not Good Enough for Canada investigates the development of Canadian immigration policy throughout the twentieth century with respect to persons with a disease or disability. With an emphasis on social history, this book examines the way the state operates through legislation to achieve its goals of self-preservation, even when such legislation contradicts state commitments to equality rights.

Looking at the ways federal politicians, mainstream media, and the judicial system have perceived persons with disabilities, specifically immigrant applicants with disabilities, this book reveals how Canadian immigration policy has systematically omitted any reference to this group, rendering them socially invisible.

VALENTINA CAPURRI is a lecturer in the Department of Geography and Environmental Studies at Ryerson University.

Not Good Enough for Canada

Canadian Public Discourse around Issues of Inadmissibility for Potential Immigrants with Diseases and/or Disabilities, 1902–2002

VALENTINA CAPURRI

UNIVERSITY OF TORONTO PRESS
Toronto Buffalo London

Toronto Buffalo London
utorontopress.com
Printed in the U.S.A.

ISBN 978-1-4875-0429-8 (cloth)
ISBN 978-1-4875-2323-7 (paper)
ISBN 978-1-4875-1985-8 (EPUB)
ISBN 978-1-4875-1984-1 (PDF)

Library and Archives Canada Cataloguing in Publication

Title: Not good enough for Canada : Canadian public discourse around issues of inadmissibility for potential immigrants with diseases and/or disabilities, 1902–2002 / Valentina Capurri.
Names: Capurri, Valentina, 1976, author.
Description: Includes bibliographical references and index.
Identifiers: Canadiana 20190177519 | ISBN 9781487523237 (softcover) | ISBN 9781487504298 (hardcover)
Subjects: LCSH: Emigration and immigration law – Canada. | LCSH: Immigrants – Health and hygiene – Canada. | LCSH: Immigrants – Canada – Social conditions. | LCSH: Canada – Emigration and immigration – History – 20th century.
Classification: LCC JV7233 .C37 2019 | DDC 304.8/71—dc23

University of Toronto Press acknowledges the financial assistance to its publishing program of the Canada Council for the Arts and the Ontario Arts Council, an Ontario government agency.

Canada Council for the Arts
Conseil des Arts du Canada

Funded by the Government of Canada
Financé par le gouvernement du Canada
Canada

Contents

NOT GOOD ENOUGH FOR CANADA

Canadian Public Discourse around Issues of Inadmissibility for Potential Immigrants with Diseases and/or Disabilities, 1902–2002

Introduction: The Personal and the Political

Penser l'immigration, c'est penser l'État.[1]

– Abdelmalek Sayad

In the introduction to *Understanding Disability*, Michael Oliver argues, "personal experience does have a direct, if complex, influence both on what gets written and the way it gets written."[2] I begin this journey by explaining how I have come to write this book and why I have written it the way I have. The idea of investigating the subject of medically inadmissible immigrants came from reflecting on my personal experience. Everything discussed in the pages that follow is filtered through the perspective of a person with multiple sclerosis (MS) who has been affected by the medical admissibility provision in Canadian immigration law. My perspective is not the Truth, nor do I simply present some factual evidence that accidently did not make it into previous scholarly works: as the late Edward W. Said wrote, facts have very little value until they get selected and incorporated into a narrative, thus acquiring context and significance.[3] This book is my attempt to articulate a narrative.

Bringing the personal in dispels the claim that scholarship can only be of significance when objective. Objectivity in academia is neither achievable nor worth pursuing. As Peake remarks, "choices about what to study and how to study are politically laden."[4] Our lived experiences and values influence our work. Claiming objectivity misguides the reader and denies our individuality. Knowledge is never neutral.[5] On the other hand, this book goes beyond writing history from a personal perspective. Writing is of little value if it does not help shape reality. Once, Howard Zinn confessed, "For me, history could only be a way of understanding and helping to change what was wrong in the world."[6] I decline to write history that only provides a window to the past. When I look at times gone by and select which facts to emphasize and how to organize them in a coherent and intelligible narrative, my goal is to affect the present I live in. Oliver explains, "the personal is the political"[7] and we can use our own lived experiences to reach an insightful understanding of, and concretely change, a reality that still oppresses persons with disabilities. In a world that keeps relying on the false dichotomy between personal and public spheres,[8] we must reassert that the two are inextricably intertwined and that our personal is never distinct from the public. We must bring our personal into the public in order to create a social, economic, and political space that addresses our needs as individuals and members of society. The focus on the personal is also a reminder that this book does not deal with numbers and statistics. It deals with human beings, each of them personally affected by decisions made elsewhere and, for those still alive, carrying the scars left by those decisions. I am one of these human beings, and here is my story.

I came to Canada from Italy in September 2001 after being accepted to the master's program in Geography at York University in Toronto. Less than a month after landing, I experienced problems with my vision and visited an optometrist. The doctor could find nothing unusual and suggested consulting a neurologist. After seeing the neurologist at St Michael's Hospital in Toronto, I was diagnosed with relapsing-remitting MS, a disabling disease of the central nervous system. I had previously experienced some symptoms of MS, but as common to many, I was told that the problem was psychological rather than physical. The day I received the diagnosis I was relieved: finally someone believed me rather than dismissing my symptoms. Little did I know that the diagnosis was two-sided: it provided me with relief, yet it also branded me for exclusion, but more about that in a moment. That same day, I volunteered to participate in a research study at St Michael's Hospital. The aim of the clinical trial was to evaluate the long-term benefits and the potential risks of a new medicine in the treatment of relapsing-remitting MS. There was no compensation for taking part in the study, and I was not charged for the medication or any research procedures, quite a bonus for a broke international student. The trial has now been completed and the new medication has been approved by Health Canada. I have remained attack-free, though my general health has shown signs of deteriorating over the years. Walking long distances is becoming difficult and running or jogging is no longer an option. I also suffer from fatigue, and when I get sick, it takes longer to recover.

In 2004, I applied for permanent residence in Canada under the independent class. As a student of Canadian immigration, I was aware that my medical condition presented an obstacle under article 38(1)(c) of the Immigration and Refugee Protection Act (IRPA). The article, known as the medical admissibility provision, establishes that applicants with a disease, disorder, or disability are inadmissible to the country if their medical condition or disability is likely to pose excessive demand on Canadian health and social services. On the other hand, each application must be considered individually: given my fairly good medical history and the fact that I have successfully integrated into Canadian society by building professional as well as personal relationships, I was hopeful the application had a chance of being accepted. The following months proved me wrong. I must give credit to the Canadian Consulate General in Detroit for trying to find grounds other than medical to reject my application. I assume that from the perspective of an immigration officer, it does not appear too offensive if the rejection is motivated by non-medical reasons. Applicants are rejected every day for a variety of different reasons. Howard Adelman claims that Canada has a responsibility to carefully select those allowed in the country since "decisions on membership in the body politics determine the future of what will become of Canada."[9] No wonder officers are cautious in the screening process. At first, I was informed that I had to pass the International English Language Testing System (IELTS) in order to receive points under the language proficiency category. This in spite of the fact that doctoral students are rarely required to complete the test. The graduate program director in the Department of History at York University kindly wrote a letter of support stating that I was able to understand, write, and speak English. Not enough. Quite a slap in the face for York University. Things are different when you have MS. I paid the fee and passed the test. I was then required to undergo a medical examination and submit a report from the neurologist at St Michael's explaining my medical history, current treatment, and future prognosis. I waited three years for a final decision on the application. Three years might seem little when written on paper but feels like eternity when your life is hanging in the balance. Applicants are reminded in correspondence that they ought to be patient since immigration officials are busy dealing with a backlog of applications, never mind that there are human beings with feelings behind their paperwork.

In May 2007, I received a "fairness" letter from the immigration officer in charge of the file informing me that I was "a person whose health condition might reasonably be expected to cause excessive demand on health or social services."[10] I was invited to submit additional information relating to this medical condition (What additional information? That I was miraculously MS-free? For those who are unaware, there is currently no cure for MS) and to indicate how I was preparing to address the issue of excessive demand. Failing to do so would result in a rejection of the application. In June 2007, I withdrew the application. I had no additional information to provide. I was also tired of fighting a lost battle, waiting for months before a final rejection. The decision to withdraw the application was facilitated by the fact that I had a choice: though Canada was the place I wanted to be, it was not my only option. I could go back to Europe and build a new life. However, different individuals have different stories and for some of them moving to Canada is a necessity rather than a choice. In some cases, acceptance can make the difference between misery and the opportunity of a lifetime. I also recognize that my situation has changed since I am now the spouse of a Canadian citizen and I did eventually apply again, this time successfully, as sponsored by my partner. In fact, sponsorship of spouses is not subject to article 38(1)(c). This time, I was therefore exempted from the provision of medical admissibility. Although the exemption has worked to my advantage, the end result was bittersweet. I wanted to be accepted for who I am, not for the person I am involved with. I am more than "the spouse of." After receiving the fairness letter, I considered the option of keeping my application independent by going to court. This route has been taken in the past, though the best result achieved to date has been allowing the claimant to remain under ministerial permit – the present temporary resident permit – without ever changing the law. The chances of winning such a battle were extremely limited. Furthermore, the financial costs involved were beyond my reach. In the country of rights and freedoms, only people with enough money are allowed to fight for their rights. The law is identical for everyone, but it offers more opportunities to those who have the money to access it.

The day I received the letter from the immigration office, I just sat down and cried. "I have determined you are a person whose health condition might reasonably be expected to cause excessive demand on health or social services":[11] reading these words on paper was a wake-up call I was unprepared for. If this book can contribute to a better understanding of the impact that the medical admissibility provision has had over time on individuals like me, it will be meaningful. If it can help in changing that reality, it might give some closure to all those who have been excluded because of a disease, disorder, or disability.

In the following pages, I investigate how and why Canada has been constructed around an idea of nation-building that has at its centre the development of a morally and physically healthy population, emphasizing a work culture that assesses people on the basis of their material contributions to the country. Adelman argues that Canada is not a charitable institution on the world stage, instead "immigrants are selected because of what Canada wants and needs."[12] His definition of usefulness fails to recognize the contributions of persons with disabilities. A capitalist system that exclusively focuses on a specific notion of work results in the erosion of disabled people's labour-power,[13] damaging their ability to assert their value as human beings and social actors. I interrogate how such a system has affected the way state and society have related to persons with disabilities by conducting a genealogical investigation to "*explain* a problematic aspect of present day history ... by examining the origins and functions of such practices."[14] I explore how the public discourse around immigration and disability entertained by selected actors – namely federal politicians, individuals, and groups whose opinions were captured by two newspapers (the *Globe and Mail* and the *Toronto Star*) – as well as the court system has changed, what has brought about these changes, what pressures were needed and who exerted them, and what still remains to be dismantled.

The book covers the period from 1902, when the Canadian medical inspection service was created as a control mechanism to prevent the entrance of undesirable foreign elements, to 2002, when the IRPA went into effect. Since 2002, no substantial changes have been registered in the existing policy. The subject of the first part of this investigation has, to a degree, been discussed by other academics and hence takes advantage of a broad range of secondary sources. Several Canadian scholars have explored various aspects of immigration with works of exceptional quality. They have helped readers understand not only who the new arrivals were but also how they transformed and were transformed by Canadian society. The analysis of those sections of the Immigration Act (in all its subsequent versions) referring to the immigration of medically inadmissible persons has instead scarcely received any attention in the academy. Addressing American scholars,[15] Longmore discusses how scholarship on immigration has failed to consider the exclusion of foreigners with disabilities in American immigration law despite the fact that, as Douglas Baynton observes, "disability is everywhere in History, once you begin looking for it."[16] In accounting for the omission, Longmore concludes that it is partly the result of the medical paradigm's dominance within society. The medical perspective has looked at disability as a problem related to the individual, therefore dismissing it as a subject for systematic study.[17] Incidentally, a similar fate seems to have befallen workers with disabilities: because they were perceived as unemployable due to their individual impairments rather than societal barriers and attitudes, Canadian working-class historiography has systematically overlooked the experiences of persons with disabilities in the labour context.[18] Though this approach has been changing in the last few decades, there is still an unfortunate lack of studies on workers with disabilities as well as on medically inadmissible immigrants. With respect to immigrant applicants with a disease or disability, Canadian immigration policy and current scholarship have rendered them socially invisible.[19] Scholars' silence on the topic could also be due to the difficulties of considering a subject that has different interpretations across time and cultures: what might be perceived as a disability at a specific time or among one ethnic group is not necessarily seen as such a few years later or among other groups. Irrespective of the cause, this persistent neglect is reflected in the paucity of books and articles on medically inadmissible migrants; hence, the second part of my research is largely based on primary sources.

The following investigation provides a major contribution to the existing literature by posing new questions on how immigration and disability have been constructed through discourse: Do immigrants and persons with disabilities belong? What are the requirements for citizenship? Why are some of us deemed more legitimate citizens than others? It also reflects on the role of politicians, the court, and sectors of civil society[20] in formulating a public discourse that validates this construction. I approach discourse as a way of speaking that is both culturally and socially organized and that identifies how a particular language was created by various social actors for talking about and understanding the topic of immigrants with disabilities.[21] At the same time, this is not exclusively a book about perceptions around immigration; rather, it is an examination of the way the state operates through legislation with the goal of self-preservation. In the process, the state uses immigration as an instrument to achieve its ends. Since the late 1860s, there has been a system in place in Canada that has shown incredible longevity and has been predicated on the belief that individuals are worthy only insofar as they are productive and useful to the material growth of the country. The exclusion of immigrant applicants with a disease or disability is the result of rooted perceptions of disabled persons as the prototype of unproductiveness. The focus on productivity has allowed the state to maintain the medical admissibility provision through time. Exclusion of medically inadmissible immigrants is presented as a tool to prevent an undue economic burden and as a necessity for the preservation of Canadian health and social services to the benefit of Canadian citizens.

In terms of organization, the book is divided into five chapters. Chapter 1 explains the methodology and terminology adopted throughout and offers insight into the situation of persons with mental and/or physical disabilities within Western, and in particular Canadian, society. The chapter discusses the notions of citizenship and social justice, exploring how citizenship has been used to exclude those perceived as other than the ideal white, male, able-bodied citizen. It touches on the concepts of power and empowerment, analysing with whom this power resides and how it is exercised. The focus then shifts to the development of Canada's immigration policy, briefly summarizing its main goals and implications for Canadians and immigrants alike. It looks at barriers and discrimination faced throughout history by potential immigrants with a disease, disorder, or disability. The two sections are connected as they both rest on the long-held assumption that an individual's value is exclusively measurable in terms of profitability.

Chapter 2 explores the public discourse developed by federal politicians. It analyses the discussions that took place among politicians in both the House of Commons and the Senate of Canada as well as in several standing committees dealing with immigration. Sources include the official report of the debates in the House of Commons and the Senate of Canada. Sessional papers containing annual reports from the Chief Medical Officers and records from relevant standing committees are also drawn on when they informed debate in Parliament. The chapter uncovers dominant perceptions existing at the federal level of government concerning the usefulness of immigration and the non-place of immigrants with disabilities. Although critical of the decisions taken and language used by several politicians, my goal is not to ridicule individuals but instead to question the system behind them. I therefore refrain from investigating the ideologies of single politicians. Focusing on individual figures is irrelevant in this context as no matter their name or their party affiliation, all were part of a system that few questioned.

Chapters 3 and 4 shift the focus from politicians to the press, investigating whether opinions expressed in Ottawa reflected and/or were reflections of understandings and perceptions among sectors of the public, namely individuals and groups who managed to have their views recorded in two major Ontario newspapers – the *Globe and Mail* and the *Toronto Star*. Chapter 3 examines the period between 1902 and the passage of the Charter of Rights and Freedoms, while chapter 4 deals with the post-Charter era. The Charter was chosen as a watershed because of its impact on Canadian society. Chapters 3 and 4 show how medical admissibility was framed in the public discourse elaborated by two major newspapers. The newspapers are a source of historical information about what some of the press was writing on the issue as well as into the assumptions of contributors and readers who expressed their opinions in interviews, editorials, articles, or letters. Newspaper analysis constitutes a narrow opening allowing the voices of social actors, at least some of them, to be recorded on matters where often only bureaucrats and politicians are heard. Here again my goal is not to focus on the opinions of specific individuals or institutions but to provide the reader with a flavour of the discourse within Canadian society. Sources consist of articles published in the *Toronto Star* and the *Globe and Mail*. The two papers started their publications on a local city scale. Over time, circulation was expanded to the wider province. Despite the *Globe*'s effort to reach out to the country, Ontario, and more so Toronto, remains the main market for both papers. It would have been interesting to examine newspapers published across Canada, particularly in the provinces of British Columbia and Quebec,[22] but expanding the analysis to cover the whole country would have required a book of its own. Toronto was selected as the principal theatre of investigation since it has historically been the main gateway city for those entering Canada.[23]

Chapter 5 examines selected legislative cases of rejected immigrants who questioned the government's actions by taking their cases to court. Legislative cases are relevant in analysing public discourse on medical admissibility because the legal system is an integral part of society and the reading of the law offered by judges is affected by mainstream understandings of disability. Court cases demonstrate that the evolving public discourse (at least in terms of language) – entertained by politicians and various members of civil society and explored in the preceding chapters – has failed to shake the ideological system at work in Canada with respect to the place of medically inadmissible immigrants. The chapter focuses on a relatively recent period of time as challenges to the law have mainly occurred since the passage of the Charter of Rights and Freedoms. Although most of the cases analysed fail to make direct reference to the Charter, its existence has given new relevance to the courts' role in the lives of ordinary individuals. Since the passage of the Charter, people have looked at the legal system as an avenue for contesting government decisions that impact their rights. In chapter 5, I reflect on how class, financial resources, and education determine who is able to take action within the legal system. My analysis indicates that there are sharp differences among those grouped together under the immigrant category. Whereas some immigrant applicants are wealthy, others can neither pay for the services of an attorney nor go through a process that is rather long and costly. These factors need to be considered in order to understand how discrimination is even harsher within certain subgroups.

The conclusion summarizes the main findings, takes stock of major changes over time, and reaffirms the argument that the provision of medical admissibility is unfair and discriminatory. The provision is the legacy of a specific historical context, yet its maintenance today seems in conflict, at least in principle, with the country's constitution. We need to ask why such an oddity has not been addressed and what role the dominant ideological system has played in maintaining the status quo. I suggest that a more profound analysis of the Charter is required in order to fully understand its implications with respects to issues of equality.

The document has a high symbolic value as the embodiment of Canadian identity. I hope that my book will raise awareness among academics around the need for further investigation of the Charter, especially its section 15(1). This section is also known as the equality clause and refers to guaranteed equality rights for every individual in Canada, also explicitly prohibiting discrimination on the basis of race, national or ethnic origin, colour, religion, sex, age, and, most important in the context of this book, mental or physical disability. The pages that follow are aimed at drawing attention to the past and current situations of foreign medically inadmissible persons trying to enter Canada. It is the condemnation of a system that for various reasons of indifference and convenience has always eluded recognizing the rights of those who do not belong.

The book is organized following a tripartite structure – Parliament, the press, the judicial system. Several events recur throughout the three sections albeit presented through shifting epistemological and documentary lenses. This structure illustrates how the three sectors follow their own discourses – each unique in the language adopted yet almost identical in its content. In the 1950 movie *Rashomon*, director Akira Kurosawa recounts the story of a rape four times, each through the eyes of a different witness, in order to show the relativity of truth. My approach is similar though the outcome is different: whereas Kurosawa explores how four different characters remember the same story in different ways, I illustrate how three separate actors attempt to recount three different stories, even though they are caught in the same narration. The repetition of certain stories is not casual but essential to grasp the overlapping of the three spheres. Whereas this tripartite structure is a significant departure from the organization adopted in a typical historical monograph (which for obvious reasons tends to configure the material chronologically), this approach is an integral part of my undertaking. If my purpose had been to exclusively narrate a series of facts explaining how immigration policies on medical admissibility have been developed in Canada, a chronological organization of the material would have sufficed. As it stands, I have struggled to achieve a different outcome: the following pages are not only about what occurred but also about public discourses developed around these occurrences. Events do not change. What does change is the analytical framework adopted by politicians, the press, and the courts in their interpretation of these events. The stories recounted here, taken together, provide the reader with an organic vision of how each piece of the puzzle is connected to the whole. Separate, they mean little. Together, they reveal a fragment of our history that has been consistently silenced.

This book is intended for a diverse audience. Within academia, students, instructors, and researchers in the fields of disability, history, and citizenship and identity in Canada will find it helpful for understanding some of the dominant societal perceptions with respect to persons with disabilities throughout the last century or so. Activists, advocates, and members of disability organizations whose mandate is to affect disability policy and politics will also benefit from this book, gaining a better understanding of the way in which the exclusion of immigrants with disabilities does not merely concern foreigners but Canadians with disabilities as well. Policy analysts, federal politicians, and federal officials should find it quite interesting to be reminded of what has occurred in the past, what has brought us here, and the reasoning behind these developments. Finally, constitutionalists and scholars of the Canadian justice system can find in the following pages inspiration and insights to reassess the role of the Charter more than three decades after its implementation.

1 The *Right* Citizen

This book examines the history of public discourse surrounding Canadian immigration policy with respect to medically inadmissible immigrants throughout the twentieth century. It investigates how the public discourse entertained by federal politicians, two major newspapers, and the judicial system has unfolded to legitimize the rejection of immigrant applicants with diseases or disabilities. The medical admissibility provision has historically allowed the Canadian state to assess who is the legitimate citizen, what are the requirements for citizenship, and what are the criteria used in establishing these requirements. In *AIDS and the National Body*, drawing from Wittgenstein, Thomas Yingling remarks, "national identity requires an ideal conception of the body and the rejection of accommodation to Otherness."[1] Women, persons with disabilities, as well as visible and sexual minorities have been marked as inadequate because they depart from the ideal of the white male citizen.[2] Catherine Holland explains that within the context of American liberalism, citizenship has been formulated as applying to the disembodied entity of the abstract citizen.[3] For persons with disabilities, this translates into society's unwillingness to take into consideration their needs and rights because they do not fit the ideal paradigm.[4] Claudia Malacrida notes how income support policies in Canada, whose goal is to provide assistance to persons with disabilities, "are written and implemented with an ideal, individual citizen in mind,"[5] someone who has the resources, both in time and money, to navigate a quite complicated and arcane set of rulings. For persons with disabilities who are also immigrants, the situation is far worse as society is not only unwilling to consider their needs but refuses to even picture them as individuals who could possibly be included within the national space. They are a non-problem insofar as their bodies remain absent from the physical and social space of the nation.

Several of the terms used in this book, such as citizen, citizenship, and disability, are historically and geographically contingent. Titchkosky explains that "what counts as an impairment changes from one culture to the next, over time, and from one person to the next."[6] The term disability is relatively new and has no universally accepted definition. In its dealings with persons with mental or physical disabilities, Canadian legislation has adopted through time what Ena Chadha describes as "assorted vocabulary."[7] In earlier versions of the Immigration Act, inadmissible immigrants were defined as lunatic, insane, and idiot. These terms were replaced in the 1960s with mentally retarded and handicapped, which were eventually discarded in favour of persons with mental and/or physical disabilities. The adoption and subsequent rejection of such terminology reveal Canadian society's changing perception of those who fall outside the normative ideal of the white, able-bodied citizen. These attitudes affect prospective as well as current citizens. I suggest that the capitalist system in Canada, particularly in its current neo-liberal version, is implicated in the discrimination faced by persons with disabilities and diseases as it is predicated on the assessment of individuals merely in terms of their economic value.[8] The emphasis on material and economic utility, far from being exclusive to Canada, is at the core of the entire Western capitalist system. In the early 1980s, Foucault had already uncovered how, throughout the development of capitalism in the eighteenth century, Western society had identified the body as the bearer of new variables such as utility, employability, profitability, and suitability to be trained.[9] The significance given the body by these new variables is at the core of Foucault's assertion that "there is a direct connection between the body and political power."[10] In fact, in order to maintain itself, political power must develop strategies and tactics whose goal is to exert dominance over the body(ies) contained within the political unit under its control.

This book questions Canada's immigration policy from 1902 to 2002, particularly the process of medical exclusion governing immigrant selection. My investigation begins in 1902 because it was in that year that the Canadian medical inspection service was set in place.[11] Whereas the first federal Immigration Act passed in 1869 included an exclusionary clause targeting paupers and disabled people,[12] it was only in 1902 that under pressure from the United States, the Canadian government amended the act by explicitly naming those suffering from dangerous and infectious diseases as among the prohibited classes. The addition was deemed necessary because, though the 1869 Immigration Act established entry standards, control mechanisms were ineffective in the years leading up to the new century.[13] Up until the 1900s, the main concern was in attracting rather than rejecting immigrants. In the initial three decades after Confederation, immigrants from Great Britain and Western Europe found the United States a more attractive place to settle.[14] Census figures indicate that throughout the same period, emigration from Canada exceeded immigration.[15] It was only after the election in 1896 of the Laurier Liberal government that immigration began its expansionary phase, thus propelling the adoption of more selective measures.[16]

Having been blessed with a steady influx of immigrants well before Canada, the United States had amended its immigration law in 1891 by providing federal officials with the authority to inspect and exclude immigrants. Following the US example, in 1902 Canada created a medical inspection service tasked with the job of excluding "any immigrant or other passenger who is suffering from any loathsome, dangerous or infectious disease or malady."[17] Replicating the US model, Canada employed medical inspection as a tool for selecting the right immigrants. Throughout 1916, rejections based on medical grounds represented over 40 per cent of total rejections.[18] The year 1902 was also the year the federal government began publishing systematic data on rejection and expulsion of medically inadmissible immigrants. In the period between 1902 and 1939, individuals deported for medical causes numbered 10,840 of the total 59,734 deportees, while 5,961 persons were refused admission for medical reasons at ocean ports of entry out of a total of 22,142 rejections.[19] Concerns with those entering the country had become a priority for the Canadian government.

While 1902 represents the starting point of the investigation, the year 2002 has been chosen as its end because it was in that year that the IRPA went into effect. Since then, the legislation concerning medically inadmissible immigrants has not been substantially modified due to the emergence of different preoccupations within Canadian society. Although the Trudeau government announced an increase in the excessive demand cost threshold in 2018, they refused to repeal the policy.[20] However, it is increasingly apparent that the policy itself is untenable. Canadian society is aging and, despite medical progress, an increase is expected in the number of persons who self-identify as disabled or ill. Canada is a country of immigration and, with growing competition from other countries, might eventually be forced to relax its admission criteria in order to attract immigrants. In addition, the impact of groups advocating for the rights of persons with disabilities is inevitably going to affect Canadian society and legislation. It looks like we have not yet reached the last chapter in the history of the medical admissibility policy.

After a brief explanation of my methodology, I present the broader picture of life in Canada for persons with disabilities, then proceed with an overview of the history of immigration to this country. In both instances, I show how individuals are constructed as valuable only insofar as they can materially contribute to the economic system. I therefore conduct an analysis of power and empowerment, asking who, in our society, has the power to make decisions and in whose interest these decisions are made. Drawing from van Dijk's work, I recognize that in order to understand power, we need to contextualize it and uncover how it is supported and legitimized by political discourse, sanctioned by the legal system, and ideologically reproduced by the media.[21] Only through uncovering these groups' access to discourse does it become possible to appreciate the way power is organized, naturalized, and institutionalized. Following Neil Boyd's argument,[22] I consider the intentions and consequences of law-making by investigating how discourse surrounding immigration and disability helps us elucidate how the law – in its formulation, application, and interpretation – plays a constitutive role in population control. I take advantage of the analysis conducted by Michel Foucault on the functioning of power as a relationship of domination that uses the law, with its institutions and apparatuses, to exercise control over different sectors of a population.[23] My goal is to show how the legislative system, far from being impartial, is among those elite institutions that contribute to the maintenance of a structure of power and oppression.

Before entering the discussion, I want to clarify the meaning given to the term disability throughout the book. Several disability rights groups oppose the inclusion of illness within the disability category. The disability movement has fought since the 1970s against the association of illness and disability, which had historically led to neglecting autonomy and the right to choose while focusing on rehabilitation at the hands of medical professionals. Associating illness and disability obscures the fact that several individuals with disabilities are healthy. While this is true, it cannot be ignored that "many people with disabilities are also ill,"[24] and many of those with illnesses are affected at some point in time by some form of disability. I myself exemplify the interconnectedness between disease and disability: as a person with a chronic disease, I have gradually experienced an increasing impairment of my abilities. In fact, I cannot run, jump, stay awake late through the night, or carry heavy loads. When asked, I self-identify as a disabled person because disability is the most relevant factor shaping my lived experience and interaction with those around me. It is disability rather than the disease that has shaped my life in the last nineteen years. Impairment is a constitutive part of the life experience for those we define as disabled, but this is not meant to deny that there are societal barriers that systematically reduce access to workplace, housing, and places of leisure and entertainment for both persons with disabilities and individuals with illnesses.[25] More important, Canadian immigration policy collapses illness and disability into the same category and gives medical practitioners the authority to assess immigrant applicants with illnesses as well as those with disabilities. For these reasons, the following pages adopt a broad definition of disability that includes illness.

It is also crucial to clarify that all references to persons with disabilities are not meant to suggest homogeneity within that category. A wide range of factors, among them gender, age, class, ethnicity, education, occupation, and immigrant status, operates to make the experience of disability unique to each and every individual. While persons with disabilities face common forms of social oppression, not everyone experiences all aspects of the oppression at all times. Parin Dossa warns that "social markers of difference cannot be dismissed under the seemingly neutral category of disability."[26] Yet, laws such as the Immigration Act continue to depict disability as a unitary and monolithic category. Despite the government claim that every individual is assessed on a case-by-case basis, the record indicates that being classified as a disabled person triggers immediate rejection. For Canadian immigration authorities, disability remains a homogeneous category that is synonymous with economic burden and unproductivity.

In this book, I've adopted an interdisciplinary approach that reflects the centrality of disability to a variety of academic fields. Concerning the discourse around disability, I have relied heavily on the literature on the social model of disability, which, since the 1980s, has become dominant in disability scholarship.[27] I have framed disability not as a problem of the individual but as a social construction that prevents persons with disabilities from integrating within society through the creation of physical barriers as well as discriminatory policies and practices. Drawing from the analysis of disability conducted by Tanya Titchkosky, I interrogate disability as a "social and political phenomenon"[28] that cannot be understood outside its specific historical and cultural context. While indebted to the social model, I also recognize its failure to integrate the body and its more or less severe impairments into disability discourse. Under the claim of their inherent biological nature, the social model excludes the body and impairment from any discussion; I instead embrace an understanding of impairment as "both an experience and a discursive construction."[29] Paraphrasing Liz Crow, it is a fact that my body is impaired and functions with difficulty, but it is just an interpretation, and one that I strenuously resist, to conclude that my body is in any way inferior or less deserving.[30] This book, therefore, is indebted both to the work of a number of scholars who have struggled to reclaim the body and impairment as essential to the political analysis of disability and its oppression and to a renewed understanding of the social model. I refer to the work of Tregaskis, Hughes, Paterson, Crow, Titchkosky, and Dossa in the following pages because they understand the social model as embracing the experience of disability as well as impairment.

My approach to narrating the history of immigration to Canada is inspired by the scholarship of three academic giants: Howard Zinn, Michel Foucault, and Edward W. Said. My primary goal in writing this book was to narrate the stories of those who have been forgotten and silenced by official history,[31] those who never made it to Canada. I could not directly reach those hopeful yet discarded applicants nor record their voices, but at least I could tell their stories and what was said about them. In so doing, I have attempted to show that there are many different ways to interpret facts.[32] Nothing of what I write comes from a position of neutrality. The past is full of facts: it is up to us to select those we deem worthy of attention and interpret them to make sense of our reality.[33] Equally important, I have tried to show how the events that unfolded were not accidents, but the result of relationships of force and power between different groups of individuals within a specific temporal and spatial context.[34] I have aimed to complicate the understanding of clear-cut immigrant categories usually found in policymakers' and academics' discourse: independent, family class, refugee, inadmissible. I have instead unpacked these categories to show the interrelationship among different aspects of people's identities, from ethnicity to gender, sexuality, income, class, and ability/disability.[35] I agree with Audra Jennings that historians, and scholars in the humanities and social sciences in general, "have more work to do to uncover the ways that disability informed discrimination based on gender, race, class, and sexuality and the ways that disability discrimination hinged on ideas about these socially constructed categories."[36] The episodes I have selected are meant to reveal these interconnections.

In the following chapters, the focus is on public discourse and how immigration and disability have been framed within the parameters allowed by the dominant discourse existing at different points in time in the political sphere, the press, and the courts. The reason I concentrate on public discourse is because "reality is constructed through language, or more broadly, through communication."[37] My approach consists in the qualitative analysis of printed material and is heavily indebted to the work previously conducted on public discourse and critical discourse analysis by Martin J. Burke, Greg Philo, Karim H. Karim, and Teun A. van Dijk. Burke defines public discourse as referring "to collections of real individuals in the social world."[38] People do not live their lives disconnected from each other, therefore it is important to examine how diverse individuals and groups discuss public issues.[39] Whereas it is crucial to recognize that there were and are "a number of conspicuous groups that remain relatively voiceless,"[40] and that often only those possessing the cultural, social, and economic means to speak and write on certain issues are able to engage in public discourse,[41] it cannot be overlooked that many of these speeches and writings "were designed to address popular audiences, and as such can be considered the political primers and lexicons of [American] public discourse."[42] Public speaking and writing express "the opinions, beliefs, and understandings of various groups over time"[43] and constitute a valuable source for comprehending how concepts such as disability, immigration, citizenship, and rights "both registered historical change and gave shape to its outcomes."[44] Public discourse reveals how changing social categories and practices expressed through the language of privileged social actors affect reality through the creation of "patterns of understanding, which people then apply in social practices."[45] In the context of this book, I use politicians' interventions, printed material, and court decisions as expression of the mainstream public discourse in Canadian society to illustrate how language and perception served to reinforce the status quo.[46]

Disability in Canadian Society

Rejection of persons with disabilities in Western societies predates modernity. Physical and intellectual fitness was considered essential to Greek society. Romans were at best uncomfortable with deformity, and Jewish culture "perceived impairments as ungodly and as the consequence of wrongdoing."[47] At the core of the rejection was the perception of impairment as threatening notions of discipline and normality, while simultaneously exposing human failure at controlling the world. This failure made disability undesirable and propelled efforts to control and hide its manifestations.[48] Around the seventeenth century, in the European context, disability lost its connection to morality and became tied to notions of utility and profit (or the lack thereof). The first institutions of confinement were created in the attempt to prevent disorders resulting from mendicancy and idleness. Foucault perceived confinement as a response to economic problems generated by idleness and concluded that "what made it necessary was an imperative of labour. Our philanthropy prefers to recognize the signs of a benevolence towards sickness where there is only a condemnation of idleness."[49] Confinement represented the solution to unemployment and begging.[50] The result was the creation of new hospitals that were institutions established to segregate economically and unproductive people.[51] At the same time, institutionalization was profitable insofar as families had to pay for the care and cure received by loved ones.[52] In the span of a few years, institutions of confinement acquired a central role in the maintenance and profitability of the capitalist system.

Western society's discrimination of persons with disabilities reached its apogee at the beginning of industrialization in the eighteenth century following a new emphasis on the concepts of "scientific rationality" and "social progress."[53] Conrad and Schneider argue that Western societies still retain the "optimism of the Enlightenment in the belief that in science and technology will be found the means for achieving good and avoiding evil."[54] Although claiming to be bias-free, science and medicine have been used as forms of social control. Far from being objective, "the very nature of medical practice involves value judgment. To call something a disease is to deem it undesirable."[55] Medicalization of deviance leads to the individualization of social problems, thus portraying disability as a universal category "untouched by political and social organization."[56] Individualization of disability results in blaming the individual's body as inadequate while hiding how society is implicated in the construction of categories such as functionality and normalcy against which the individual body is assessed.[57]

Within Western tradition, disability has been trapped in a rhetorical discourse of cost and managerialism, and persons with disabilities have been perceived as unproductive and burdensome to the public purse. Disabled persons are those who, in the words of Parin Dossa, "cannot give but only receive."[58] By the end of the nineteenth century, industrialization had brought to the fore a work ethic that translated in "horror of chronic dependency."[59] Industrial capitalism devalued the labour-power of disabled workers and used disability to exclude those workers "from the principal means of survival in a wage labour economy."[60] In the mid-1800s, Samuel Gridley Howe, American teacher to the "feeble-minded," was convinced that "this class of persons ... are dead weights upon material prosperity of the State."[61] Dustin Galer notes that in Victorian Ontario, the dominant assumption was that disabled workers were unproductive and unable to be employed in most industrial jobs, so much so that even mutual help initiatives like fraternal insurance refused to extend membership to applicants with pre-existing disabilities as they were considered liabilities and a drain on resources.[62] Between 1890 and 1920, American reformers identified "crippledom" as a problem with economic and social outcomes, and "cripples" were referred to as "parasites feeding off the labor of others."[63] Throughout the first half of the twentieth century, American society perceived "disabled Americans as dependent"[64] and therefore unable to be either wives/mothers or fathers/breadwinners, roles that were considered essential to the fulfilment of the individual as a rightful member of society.

The English and American response to dependency (especially in the case of persons with mental disabilities) was institutionalization, and the model was soon adopted in Upper Canada where the first provisional lunatic asylum was established in Toronto in 1841, to be replaced by a permanent provincial asylum in 1850.[65] Ontario's mental health system became part of a larger process of social control in the province. Although its development was peculiar to the Canadian context, "institutionalization of specific problem populations was not unique to Ontario society, and indeed the developments in Ontario were significantly affected by both English and American experiences."[66] The creation of institutions for the insane was motivated by "a concern on the part of reformers for 'effective' and 'economical' means for dealing with the deviant and dependent"[67] rather than benevolence. Patients in the asylum were regularly put to work not only because work was considered therapeutic but also because their labour was expected to offset some of the costs of the institution's maintenance.[68] As economically efficient and self-sufficient institutions, asylums were perceived as the perfect solution to the problem of human imperfection.

The establishment of mental institutions across Ontario advanced steadily throughout the 1850s with the opening of branch asylums at the University of Toronto (1856), Fort Malden in Amherstburg (1859), and Orillia (1861).[69] As the number of asylums in the province grew, so did the practice of patient labour: while in the late 1870s, approximately one-third of the inmate population was employed, the figure rose to three-quarters by 1900. Despite doctors' and inspectors' claims that work was not meant to exploit but help patients through "a benign form of moral therapy,"[70] annual reports indicate that patients' unpaid labour contributed considerably to reduce expenditures.[71] Under the excuse of providing moral treatment, asylum officials exploited patients for internal economic necessity. Records show that institutionalized individuals were put to work in asylums all over the province, though the practice was common beyond Ontario and Canada. Reaume observes that "evidence of patients' unpaid labour that contributed significantly to keeping provincial asylums operating clearly undermines broad statements about their supposed unreliability."[72] Mentally disabled persons' unproductivity was a pretext for taking advantage of the free labour provided by individuals unable to question their unpaid status and reclaim the entitlement of a wage for work.

Contemporary capitalism reveals a legacy of discriminatory industrial labour practices that continue to devalue the labour-power of persons with disabilities. As Trent notes, persons with disabilities have been exploited to the advantage of the capitalist system: in times of prosperity, their work is required, while during periods of unemployment, they are deemed unproductive and excluded from the labour market.[73] According to Barnes and Mercer, "Throughout the 20th century, the level of exclusion from paid employment exhibits some extraordinary short-term fluctuations."[74] A typical case was the drafting into the labour force of disabled persons in Britain (the same occurred in Canada) during the Second World War, and their subsequent ejection at the end of the conflict when employment became once again the preserve of able-bodied members of society. Oliver maintains that Western capitalist societies remain among the main theatres of disability oppression due to their celebration of productivity and reliance on the ideal worker as a healthy, industrious, and exploitable subject. The understanding of paid employment as a precondition for social inclusion, combined with the ongoing marginalization of those with disabilities within the labour market, has resulted in disabled persons' systematic exclusion.[75] The economic and cultural devaluation of persons with disabilities results in both physical inaccessibility and socio-spatial exclusion.[76] While this discrimination is ingrained "in the institutionalized practices of society,"[77] its underlying assumptions are questionable. Despite common misconceptions portraying persons with disabilities as "unemployable, unproductive and extremely costly,"[78] a majority of them do in fact work.[79] Those who do not work are excluded from employment by societal barriers rather than impairment. Equally important, persons with disabilities are not only workers but also consumers, as the growing market for aid and equipment for disabled people can attest.[80] Furthermore, Dossa notes that, paradoxically, persons with disabilities are blamed for their dependency while, at the same time, being forced into that dependency "in order to maintain a service industry for economic growth."[81]

By devaluing disability, society curtails the opportunities of persons with disabilities to enjoy full citizenship rights in their community of belonging. Due to the prominence given to paid work over other forms of work and other contributions to society, persons with disabilities have been denied those citizenship rights their able-bodied counterpart take for granted.[82] This is evident in Canada at the national, provincial, and municipal levels. Throughout the last three decades, changes to state policy at the federal and provincial levels have negatively affected Canadians with disabilities. In 1996, the then Liberal government in Ottawa receded from a national strategy for full integration of persons with disabilities by reducing its financial support and decreasing the number of those who qualified for the disability tax credit. Vera Chouinard points out that in Canada and other capitalist societies, the term dis-abled is increasingly used to identify those who are "negatively different from the able-bodied ideal."[83] The situation of persons with disabilities worsened under the Conservative Harper government. Despite the advocacy of organizations such as the Council of Canadians with Disabilities (CCD), "Canadians with disabilities are almost twice as likely as nondisabled Canadians to live in poverty"[84] and the same ratio applies to their chances of being unemployed or underemployed.[85] According to Statistics Canada, two million disabled Canadians are unable to access basic public services and decent jobs.[86]

Matters are made worse by the current historical conjuncture dominated by a neo-liberal political approach that relies on increasing calls for austerity and a decreasing commitment to welfare state social policies.[87] In what Chouinard defines as "an extremely ableist regulatory regime," Canada marginalizes persons with disabilities by restricting their opportunities and disciplining their demand on state resources.[88] In her study of income support policy in Canada and the UK, Claudia Malacrida points out that "particularly in the Canadian context, the policy is written in a way that constructs potential recipients as dishonest adversaries who will require surveillance and supervision."[89] Therefore, the state remains a terrain of struggle for persons with disabilities, particularly those who lack financial or political support. While capitalist society has a work culture that grants social status on the basis of people's ability to work, the definition of work is quite limited, failing to acknowledge persons with disabilities as productive subjects. Social factors such as the absence of opportunities and services as well as blatant discrimination deprive persons with disabilities of the right to employment.[90] Their labour-power is devalued and this devaluation is then blamed on the individual and used to justify further marginalization. In Canada, a whole range of socio-spatial practices of exclusion force Canadians with disabilities into the position of second-class citizens whose rights are fundamentally challenged.

In adopting productivity as the exclusive criteria to assess individuals, Western capitalist society is losing sight of the inherent value of human life. We were there not so long ago, and the record is far from enlightening. At the end of the nineteenth century, scholars writing in the British *Westminster Review* suggested that society could be relieved of its financial burden by eliminating the insane. In the following century, euthanasia, while advocated by some as guaranteeing the right to die without sufferance or use of extraordinary measures, was proposed by others as a way to relieve society of the financial burden of those lives considered useless to the community.[91] In the United States, Charles Eliot Norton, professor and former president of Harvard University, "advocated for the 'painless destruction' of insane and deficient minds."[92] When in 1935

Nazi medical philosopher Fritz Bartel invited society to rid itself of defectives,[93] he was far from a lone wolf: a Gallop poll conducted in the United States in 1937 indicated that 45 per cent of the sampled population was in favour of eliminating defective infants.[94] All throughout the following decade, several American, Canadian, and European scientists and academics supported euthanasia as a solution meant to reduce medical costs.[95] At Nuremberg, the allied prosecutors carefully avoided dealing with the Nazi extermination of persons with disabilities. In one of the few references available, Chief US Counsel Robert H. Jackson commented that disabled persons put to death "were substantial burden to society, and life was probably of little comfort to them."[96] A similar rhetoric can be found in current debates about euthanasia and assisted suicide to relieve persons with disabilities "of their burdensome lives."[97] Likewise, government documents often refer to the elderly as well as those with disabilities as a burden on society. Modern genetic testing and screening programs are presenting increasing numbers of expecting parents with the opportunity to abort fetuses identified as abnormal.[98] While parents are not forced to abort disabled fetuses, the societal pressure put on them is high and suggests that choosing to keep a disabled child will negatively impact the entire population because of the eventual services and associated costs the person will require.

The refusal to acknowledge the contributions of persons with disabilities results in paradoxical situations like the one described by Dossa whereby we can terminate the lives of fetuses that are deemed defective but at the same time maintain that persons with disabilities are worthy of respect, all without perceiving a contradiction.[99] By hiding the inconsistency in our social thinking about disability, we hand out a social death sentence to those who are deprived of any "positive social status in the wider society."[100] While persons with disabilities are the primary targets of this exclusion, the larger society suffers as well from the loss of "participation, resourcefulness and creativity"[101] of disabled individuals. Disabled people contribute to a richer and more diverse community by offering a new perspective to look at the world. In refusing to engage with and nurture the alterity embodied by persons with disabilities, we end up creating what Margaret Lock defines as "a bland world of sameness."[102]

Disability and Canadian Immigration Policy

In this section, I discuss Canadian immigration policy vis-à-vis immigrant applicants with disabilities. Taken together, this and the previous section reveal that whether the focus is on Canadians or immigrants with disabilities, the underlying dominant assumption remains that human beings are valuable only insofar as they are economically profitable. Albeit with differences in outcome – Canadians have a right to stay while immigrants are rejected or deported – both groups are assessed from a similar (though, as it will be discussed later on, far from identical) perspective. Immigration control "has been one of the major social policy instruments employed by capitalist states" to "regulate the size and character of national populations."[103] Since Confederation, Canada's ideal citizenship has been embodied in the "white, male, productive, responsible and compliant"[104] subject. Authorities took upon themselves the task of disciplining their citizens at home while rejecting at ports of entry immigrants who failed to meet the required standards.[105] While the tools adopted were different, the final goal was certainly not.

Canadian immigration law encourages immigration as necessary to the country's prosperity, while at the same time discouraging entrance of unfit subjects who are considered a burden rather than an asset. In 1869, the first Immigration Act empowered the federal government "to deny entry to paupers and the mentally and physically disabled."[106] The act categorized mentally infirm persons as either lunatic or idiotic: the former included those "considered to be of unsound mind, mentally ill, or insane" while the latter "referred to individuals with cognitive impairment, low intelligence, or developmental delay."[107] Prime Minister John A. Macdonald believed that his dream of bringing the West into Confederation could only be achieved by promoting immigration of white, Anglo-Saxon, healthy individuals. In order to create a nation out of a bunch of ex-colonies, Canada needed economic growth and a large population that only sustained immigration could have assured.[108] Meanwhile, under the influence of the hygienist movement, the government began pursuing the ideal of a national community that was genetically sane and strong, therefore categorically barring entry to those bringing the threat of degeneracy.[109]

The emphasis on strength, productivity, and sanity resonates throughout the entire history of Canada, particularly in reference to its immigrant population. In 1891, the patriotic rhetoric of the general election rested on the image of ideal citizens who were "strong, loyal, white, heterosexual, providers and protectors."[110] At the beginning of the twentieth century, Clifford Sifton, minister of the interior in the Laurier government, believed that Canadian prosperity could only be achieved with massive immigration of strong farmers.[111] His successor, Frank Oliver, confronted with growing mainstream anxieties about foreigners, stressed the necessity for British, West European, and American influxes to preserve the character of the population.[112] In all cases, immigrants were exclusively assessed according to their potential contributions to

Canada's prosperity. Martin Pâquet observes that economic utility has traditionally been among the main criteria of inclusion/exclusion adopted by the Canadian state.[113] The truth of this observation will become apparent in the following chapters. And yet, the question I am more interested in is why certain subjects are constructed as excludable in the first place.[114] Interrogating how capitalist societies are organized around a framework that allows the possibility of exclusion is important insofar as it switches the focus of attention from the excluded subject to the excluder, thus trying to understand on what grounds the exclusion is legitimized. By focusing on public discourse, I attempt to elucidate the rationale for considering foreigners with disabilities as excludable. I hope that the narrative presented will shed light on the ways certain subjects are constructed as unwanted and threatening to the national body.

The Immigration Act was first amended in 1902 with a provision for the rejection of those "suffering from any loathsome, dangerous or infectious disease or malady."[115] The amendment was the "first explicit health-related prohibition outright barring certain classes of persons from entering into Canada."[116] This was followed in 1906 by a revised Immigration Act further restricting admission criteria and naming a broad spectrum of undesirable immigrants who were rejected on the basis of their inability to fully integrate and contribute to society.[117] Among those inadmissible were individuals who were "epileptic ... dumb ... or suffering from any disease or injury."[118] The addition of the terms epileptic and dumb was meant to keep the legislation up to date with developments in the booming fields of psychiatry and eugenics. Also inadmissible were feeble-minded, including those with mental retardation as well as persons affected by poverty, alcoholism, and juvenile delinquency. In 1910, the list was further expanded when the category of imbeciles appeared in the new Immigration Act to identify persons with an IQ (a measure invented in 1904 by psychologist Alfred Binet) lower than average. The 1910 act barred mentally deficient immigrant applicants from entering the country even when accompanied by family willing and able to financially support them, something that had been tolerated in the previous version.[119] Several scholars supported political efforts at excluding undesirable migrants. In *Strangers within Our Gates*, first published in 1908, James Shaver Woodsworth wrote that the time had come to deal with immigration and "shut down those whose presence will not make for the welfare of our national life."[120] In 1922, William G. Smith invited Canada to reject immigrants who lacked "the health ... to tackle the tasks of agriculture, which ... must be our basic industry."[121] Child immigration was only allowed for children with the potential to become self-supporting citizens. In Ontario, the 1897 Immigrant Children Act established that only "children of sound physical condition"[122] were admissible. Quebec, Manitoba, and Nova Scotia soon followed with similar legislation.[123]

By the 1920s, intolerance of undesirable immigrants was entrenched in a number of institutions. In a 1921 report, the Canadian National Committee for Mental Hygiene called on the federal government to "so guard our ports of entry that we do not receive an undue proportion of those who will become eventually a burden to the State."[124] An editorial published in 1920 in the *Canadian Journal of Mental Hygiene* noted that a large portion of the mentally disabled in the province of Manitoba were immigrants.[125] In her study of mental institutions within the province of Alberta, Malacrida notes, "concerns about the troubling nature of both Eastern Europeans and First Nations People were clearly expressed in eugenics discourse in North America during the early part of the twentieth century."[126] In response to such concerns, federal politicians passed the consolidated 1927 Immigration Act, adding persons of constitutional psychopathic inferiority to the ever-growing list of inadmissible categories.[127] Following the Second World War, in the middle of an industrial transition and with increasing demand for workers, the desire for a selective immigration policy remained a priority as demonstrated by Prime Minister Mackenzie King's statement in 1947 that "Canada is perfectly within her rights in selecting the persons whom we regard as desirable future citizens."[128] Built on this guiding principle, the Immigration Act of 1952 thoroughly listed all the prohibited classes and established the requirement of medical examination for immigrants.[129] The list of inadmissible persons included idiots; imbeciles; the insane; those with constitutional psychopathic personalities; epileptics; those with tuberculosis or any other infectious disease; the dumb, blind, or otherwise physically defective (unless they had sufficient means of support); and all those certified by a medical officer as being mentally or physically abnormal. The act also declared inadmissible all individuals of a family when at least one member had been assessed as mentally disabled.[130]

The situation hardly improved under the short Conservative interregnum from 1957 to 1963. With serious economic problems and high unemployment rates, it was a tough period for immigration. In 1962, the new regulations introduced by Minister of Immigration and Citizenship Ellen Fairclough began the process of replacing race with skill as the main criterion of immigrant selection yet did nothing to address discrimination against immigrant applicants with disabilities. Five years later, with the introduction of the point system, age, education, skills, language capability in one of the official languages, and degree of kinship with relatives already in Canada became the criteria used in assessing immigrants, thus masking the still racist and gendered nature of immigration policy.[131]

While the country allegedly left behind its history of racial discrimination, immigration continued to be a tool in support of Canada's interests by becoming another "aspect of the employment market."[132] The 1976 Immigration Act, the jewel in the crown of the Liberal government of Pierre E. Trudeau, did not represent a departure from previous policy concerning inadmissibility of immigrant applicants with disabilities. The alleged end of racial discrimination in immigration policy left untouched the long-standing discourse about immigrants' utility to the country. In this sense, the point system solidified Canadians' understanding of immigration as a vehicle serving national economic development.[133] Within this understanding, foreigners with disabilities continue to be perceived as undesirable.

In this chapter, I have provided a brief overview of the situation experienced by persons with disabilities in the Western world, with a particular focus on Canada, a country built on a capitalist system that rests on the link between bodies and productivity. Within such a system, persons with disabilities are portrayed as unfit and burdensome. Canada's immigration policy has been developed around the notions of usefulness and productivity. The legislation has been designed to attract and retain able-bodied subjects while discarding those failing to fit the requirements, among them persons with disabilities. The result has been a selection process that has seen innumerable lives sent back and forth across the oceans because they were considered expendable. As a matter of fact, non-Canadians have historically been forbidden from claiming the status of disabled within the Canadian national community since they have been systematically stopped at the borders. In the following chapters, I explore how the exclusion was legitimized in the public discourse formulated by dominant institutional actors – namely federal politicians, the mainstream press in Ontario, and the courts – and how this rejection has resulted in the discrimination and oppression of those who, in the eyes of society, do not fit in. It is a testament to people who have been ignored, a story that has waited a long time to be told.

2 Parliament and Medically Inadmissible Immigrants

According to article 38(1)(c) of Bill C-11, the Immigration and Refugee Protection Act that received royal assent on 1 November 2001, "A foreign national is inadmissible on health grounds if his/her health condition might reasonably be expected to cause excessive demand on health or social services." The subsequent paragraph clarifies that the clause is not applicable to a foreign national who:

(a) has been determined to be a member of the family class and to be the spouse, common-law partner or child of a sponsor within the meaning of the regulation;
(b) has applied for a permanent residence visa as a Convention refugee or a person in similar circumstances; is a protected person; or
(c) is, where prescribed by the regulations, the spouse, common-law partner, child or other family member or a foreign national referred to in any of paragraphs (a) to (c).[1]

The provision of medical admissibility predates 2001; though worded differently and with some variations in its content, it has been present since the beginnings of Canadian immigration policy. Whereas the original intention was to exclude subjects who were considered a risk to the genetic integrity of the nation or unemployable, over time more emphasis has been placed on the excessive cost that the disease or disability would likely add to Canadian medical and social services. The concern for what the immigrant can do for the country has been replaced with what the country is not willing to do for the immigrant. Both perspectives are predicated upon the belief that immigrants' worth ought to be measured in terms of usefulness to the receiving country. In a nutshell, Canada wants individuals who will contribute to its economic growth while using its services as minimally as possible.

This chapter examines how the provision of medical admissibility was justified in the public discourse framed by federal politicians. Despite the overall continuity in approach, three consecutive phases are noticeable. The years immediately following 1902 were characterized by general uneasiness towards all persons with mental or physical disabilities, whether citizens or immigrants. Disabled persons were considered a burden. While immigrants could be excluded, disabled citizens had to be accepted as a misfortune, although repeated efforts at passing sterilization legislation were made at different points in time.[2] The late 1940s and early 1950s witnessed a shift to the second phase as the impact of returning disabled veterans coupled with the advancement of disabled citizens' rights groups brought new awareness to the contributions Canadians with disabilities were making in the life of the nation. Immigrants with disabilities continued to be rejected, but the language justifying the exclusion was refined in more acceptable terms. This is true for the terminology used in the legislation as well as the language adopted in Parliament. A final shift occurred in the mid-1980s, after the passage of the Charter of Rights and Freedoms. The clause of non-discrimination against, among others, persons with mental or physical disabilities represented a major legal boost to the rights of Canadians as well as immigrants, especially after the Supreme Court established that everyone in Canada, independent of legal status, is entitled to the full protection of the Charter.

At the beginning of the investigation, my foremost concern was uncovering what motivated Canadian parliamentarians to pass and maintain legislation concerning immigration of individuals with diseases or disabilities. Seeking an answer, I plunged myself in pages and pages of debates that took place from 1902 to 2002 in the House of Commons and the Senate of Canada. Parliamentary debates were chosen as the main documentation because, as Patricia Roy explains, "the published debates of the House of Commons and, to a lesser extent, the Senate are a very useful and easily accessible source of political opinion."[3] Whereas through time many of the policies passed in Parliament were directly or indirectly pushed forward by the Department of Immigration, it remains true that "politicians made immigration policy"[4] and they were ultimately accountable for it. Critics might argue that my analysis fails to consider that since the last half of the twentieth century, decisions have increasingly been made in the way of regulations issued by the department rather than legislation passed in Parliament. True enough, though I maintain that it has been Parliament's collective choice to abdicate its power and responsibilities. The legislative power in Canada rests with Parliament.[5] Whether bureaucrats have a role, and they certainly do, is inconsequential in the context of this book: rather than understanding who prepared immigration bills, my goal is to uncover why a collective of members in the legislature agreed to pass them into laws and formulated a public discourse that justified and legitimized the language and content of those bills.

This chapter presents a sample of parliamentary discussions and is not intended to provide an exhaustive historical summary of pertinent bills debated in the House of Commons and Senate. My concern is with selecting some of the interventions made by members of Parliament (MPs) and senators in order to examine the language used through time when referring to persons with a disease or disability, also showing how that language was shared, with few exceptions, by members of all political parties. I am not concerned with the specific views of individual politicians; it is irrelevant who made a certain comment, as every one of them was part of a structure they all accepted and worked to preserve. My goal is to show that there was a system in place rather than speculating on the reasons the provision went uncontested by single MPs or senators. Looking at the first three decades of the twentieth century, Martin Pâquet observes that despite their ideological differences, Canadian parliamentarians agreed that the ideal immigrant had to bring an economic contribution to the host country.[6] That belief persisted in the following decades. My effort in this chapter is therefore meant to uncover the rationale behind such collective assumption.

Phase One: 1902 to Early 1940s

Let's then start a journey back in time to encounter some of these politicians. It was over a hundred years ago, on 16 April 1902, that Clifford Sifton, minister of the interior in the Liberal government of Wilfrid Laurier, introduced Bill (n.112) to amend the Immigration Act denying entry to those with "any dangerous or infectious disease."[7] Sifton was among the supporters of an immigration policy aimed at bringing strong and healthy farmers who could contribute to the country's economic growth.[8] During the debate following the introduction of the bill, on 29 April 1902, Mr Edward F. Clarke (Conservative, West Toronto) called for more stringent admission requirements. In his intervention, Mr Clarke, ultra-Protestant Orangeman and former mayor of Toronto, complained, "It is a lamentable fact ... that a large number of diseased immigrants have been brought into Canada, are now in Canada, and are being maintained at the expense of the Canadian people."[9] The issue had international ramifications as the United States was increasingly exerting pressure to halt entrance of diseased or disabled persons into Canada. On 14 July 1903, in the Committee of Supply of the House of Commons, Mr Uriah Wilson (Conservative, Lennox, Ontario) quoted Mr Watchorn, US commissioner of immigration, as reporting: "The Canadian route to the United States is known to every unscrupulous agent in Europe, and is by that means made known to the very dregs of society, many of whom having been rejected at United States ports sought this easy mode of escaping the effect of official vigilance."[10] Blaming Canada for lack of border control touched a nerve with parliamentarians. The opposition accused the government of a lax attitude while members of the cabinet resented the unsubstantiated attack. Sifton responded, "We consider that the officers of the Canadian government are just as competent as those of the United States government."[11] Obviously, the accusation hit a nerve.

Immigrants' medical inspection and border crossing into the United States remained hot topics throughout the following months. Despite Sifton's remarks that "at the beginning of the year we had inaugurated a system of inspection which I have reason to believe is about as perfect as we can make it,"[12] the debate was far from over. Mr Clarke questioned the government: "Year after year there is published a statement showing the number of these immigrants from Europe wending their way to the United States through Canada, who are prohibited from continuing their journey and who are left stranded in Canada. What is to be done with these people? Can there not be an improvement made in the medical examination?" The matter was of importance because the purpose of immigration was fostering Canada's development. Immigrants were expected to be hardworking and productive people who contributed to the growth of the nation. In the Senate, Hon Mr John Henry Wilson (Liberal, Ontario) remarked, "the government ought to ... do everything in their power to prevent the admission into this country of any persons whose presence will not be of lasting benefit to the Dominion."[13] In 1905, Mr John Hoolahan, Montreal agent, wrote in his report to the superintendent of immigration that "persons suffering from mental or physical disability or aged people, are not desirable addition to the population of Canada."[14] He added, "Canada is too young and vigorous to welcome any immigrant who is not self-sustaining."[15] On 10 May 1909, during discussion around Bill (No.17) respecting immigration, Mr Eugene Paquet (Conservative, L'Islet, Quebec) explained who was the ideal immigrant: "We want here desirable immigrants, physically and mentally sound, in order to till our fertile lands ... and also to develop our agricultural industry, our lumbering industry and our mineral wealth. On the other hand, we wish to eliminate that emigration which does not contribute to the material and moral progress of the Canadian people."[16] The government was invited to exercise a stricter control over the admission of immigrants since, as Hon Mr Henry J. Cloran (Liberal, Quebec) noted in Senate, "It is all very fine to bring in large numbers of people, but it is better to ensure quality."[17] But exactly who should have been kept out? According to the Instruction for the Medical Inspection of Immigrants contained in the Annual Report of the Department of the Interior for the Fiscal Year Ended 30th June, 1904, there were three classes of people who "may not be admitted into Canada," or only admitted under certain conditions:

- Class I. Those persons who by reason of some specified physical disability or disease, or through some moral or criminal cause, are refused admission to Canada.
- Class II. Those persons who by reason of being diseased, crippled or deformed, or through some mental condition, must be held for examination as to whether the conclusion "that they are likely to become a public charge" can be justified.
- Class III. Those persons who, suffering from some physical disease of a curable character, may be admitted for treatment to a Detention Hospital under the supervision of the department, under the conditions set forth in the Act or Regulations.[18]

The report further elaborated that the term physical disability pertained to "insane persons, epileptics, idiots, blind, deaf and dumb persons and other defectives," quite a wide spectrum of anomalies.

Confusion among politicians around who was to be allowed in resulted from applying an old act to a situation that, due to the increase in immigration throughout the decade, required an updated set of provisions. In 1906, Bill (No.170), An Act Respecting Immigration and Immigrants, was introduced with the purpose of amending the law in order to make it suit modern requirements. Some of the new sections were controversial and were the subject of long debates in both Houses of Parliament. Special attention was given to section 28 that read: "No immigrant shall be landed in Canada who is feeble minded, an idiot, or an epileptic, or who is insane, or has had two or more attacks of insanity within five years; no shall any immigrant be so landed who is deaf and dumb, blind or infirm, unless he belongs to a family who accompany him or are already in Canada and who give security, satisfactory to the minister ... for his permanent support if admitted into Canada."[19] Among the most debated points was whether the rejection needed to be strictly applied or should some discretionary leverage be granted the minister in evaluating the situation on a case-by-case basis. Several politicians were of the opinion that not all situations called for exclusion from Canada and that many persons with a physical infirmity could function well in society. This was (and still is) in line with the traditional hierarchical ranking of persons with disability within the Western context; as noted by Parin Dossa, it has been customary to consider physical disabilities as less severe and therefore more acceptable than other forms of disabilities.[20] On 13 June 1906, in the House of Commons, Mr Guthrie (Liberal, Wellington South, Ontario) wondered, "Would the minister be prepared to exclude a man merely because he was deaf?"[21] Mr Haughton Lennox (Conservative, Simcoe South, Ontario) echoed these concerns and noted, "Deafness should hardly exclude an immigrant from the country."[22] The issue was also discussed in Senate where Hon Mr John Henry Wilson (Liberal, Ontario) observed: "As far as epileptics are concerned ... we have any number of individuals so afflicted in Canada who are good citizens and are making a good living not only for themselves but for their families. We are well aware that many of those who are deaf and dumb are good citizens. They marry and intermarry and carry on their vocation, farming or whatever it may be, and are useful citizens in performing their ordinary duties of life."[23] Despite similar arguments voiced in the Senate, in the end the majority voted to retain the exclusion for all immigrants who fell under section 28.

A related concern was whether persons with disabilities whose family could assume the cost should be accepted, particularly when they were minors. Hon Mr Lawrence Geoffrey Power (Liberal, Nova Scotia) commented, "If a father and mother come to this country bringing with them say half-dozen children and one of those children happens to be feeble-minded or an epileptic, it seems to me that he should be allowed to come in ... If the afflicted person ... has a family who are prepared to take care of him, and to give security for his permanent support, then he should be admitted."[24] Some senators were sympathetic, though the general consensus was that Canada's interest had priority and strict admission requirements were a necessity, provided the minister was left with enough discretion to accommodate individual situations. Hon Sir Richard Cartwright (Liberal, Ontario) remarked, "there has been more or less a systematic attempt on the part of certain authorities in the old country and elsewhere to dump on Canada persons who would be a charge."[25] Once the government was given powers to prevent the entry or deport undesirable immigrants, Hon Mr Richard William Scott (Liberal, Ontario) reassured his colleagues that exceptions for humanitarian reasons were going to be made as "you must have some sort of confidence in the government administering the law in a humane and proper manner."[26]

A further problem was deciding on the length of time allowed to deport individuals who had been accepted but subsequently developed any "unfavourable characteristics." Bill (No.170), the first piece of legislation to legalize deportations in Canada, provided that immigrants could be deported up to a period of two years after their arrival.[27] According to Mr Oliver, minister of the interior, the provision was justified because of the initial difficulty in "detecting undesirable conditions."[28] Mr F.D. Monk (Conservative, Jacques Cartier, Quebec) explained that "there is no examination at present that will ascertain if a man is feeble minded, or epileptic or if he has attacks of insanity."[29] A few MPs pointed out that the provision was ethically questionable since, in the words of Mr Lennox: "it is contrary to all natural justice that a person who comes into Canada ... in perfect health ... and who, within two years afterwards, perhaps without any fault of his own, becomes a charge upon the public funds, municipal, provincial, federal or otherwise, should be deported in consequence."[30] Others, including Mr Monk, believed in the necessity of protecting Canadian institutions from foreign elements threatening to exploit and impoverish the country: "We have a certain number of immigrants who are sent out here by their families, some of whom may be epileptic, others with diseases that were not apparent when they landed, and they are sent out here for the purpose of getting rid of them. After they have been here a certain time they end up by getting into our hospitals, our asylums or our jails."[31] Similar divisions were registered in the Senate where Hon Mr James Drummond McGregor (Liberal, Nova Scotia) was inclined to allow a "worthy citizen" who had become disabled to stay, while Hon Mr Robert Watson (Liberal, Manitoba) maintained, "it is very important to have a provision whereby such persons could be deported instead of allowing them to remain as a charge upon the community."[32] Despite Hon Mr McGregor's attempt to pass an amendment reducing the time allowed for deportation from two years to six months after arrival, the amendment was declared lost on a division.

The debate would not subside and on 3 April 1907, during discussion in the House of Commons around Bill (No.143) introducing an amendment to the Immigration Act that would have made operational Bill (No.170) passed the previous year, Mr Foster (Conservative, King's, New Brunswick) reiterated, "Now a man comes in who is perfectly healthy, all his antecedents are good; he remains in Canada for the first year, and some time during the second year a misfortune happens to him ... On what proper ground can you deport that man?"[33] As elected representatives, politicians had a responsibility to take into consideration public opinion, and in this case public opinion meant voters. Mr Oliver noted that "there is a very strong demand on the part of public opinion for the prevention of undesirable immigration, and if it is desirable to prevent it, and we have taken every means we can to prevent it, but still the undesirable is found in our midst, public opinion seems to demand that deportation should follow."[34] Hardly surprising, the bill passed.

A similar orientation is detectable in the discussion taking place three years later, when the House gave second reading to Bill (No.102), another piece of legislation intended to amend the Immigration Act, further extending the time for deportation. If anything, opposition to immigration and calls for deportation of diseased and disabled immigrants were even stronger due to the economic crisis that hit the country following the 1908–9 depression.[35] Mr Uriah Wilson (Conservative, Lennox, Ontario) explained: "The minister has increased the time for deportation from two to three years. Dr. Clarke, of the insane hospital of Toronto, is strongly in favour of increasing the period to four years. For myself I can see no reason why there should be any limit fixed to the time within which deportation may take place. If people come into this country who are not able to take care of themselves, who have to be sent to an insane asylum, or other places, to be supported by the public, I do not see why we should retain them at all."[36]

Addressing his colleagues, he appealed to their patriotism and recalled their obligation to protect future generations: "That is a question in which not only we are personally interested, but in which our children and our grandchildren will be still more interested, because they will have to live side by side with these undesirable people who succeed in entering our country. We ought to take every precaution to keep the physical standard of Canadians as high as possible."[37] Commiseration was misplaced since most of those people "are sent to this country by relatives who are anxious to get rid of them."[38] Apparently, the ones who should have been pitied were Canadians, not deported immigrants.

At this point, curiosity demands a brief look at the numbers of deportation. The Report of the Chief Medical Officer for the Fiscal Year 1908–9, included in the Annual Report of the Department of the Interior, indicates that, among other individuals, 15 persons were deported for rheumatism, 113 because insane, 35 because feeble-minded, 22 for epilepsy, 11 because crippled, 82 for physical debility, and 14 for physical and mental debility.[39] The report established a connection between insanity and specific nationalities. Its author, Dr Peter H. Bryce, chief medical officer from 1904 to 1921, was surprised to find out, in analysing the collected data, that persons from southern countries showed few cases of insanity. This was contrary to Dr Bryce's belief that non-Anglo-Saxon races were prone to infirmity and insanity.[40] The report observed: "The notable absence of mental defectives amongst the people from southern countries is a matter of much interest and, contrary to a too popular opinion, it appears that ... we have in such races not only an industrial asset of great value but also the assurance of a population remarkably free from the degenerative effects seen in those classes which have been for several generations factory operatives and dwellers in the congested centres of large industrial population."[41] In 1914, Dr Bryce reiterated: "It has been remarked in previous reports that very few Italians and equally few Orientals become insane, and ... pauper inmates of hospitals ... the fact in both cases that these ruder foreigners are almost all manual labourers, means that as outdoor workers they are naturally healthy."[42] What a fortunate coincidence that those people were just the kind Canada's economy needed so badly at the time! Notably, Bryce never referred to immigrants from Asia or southern Europe as prospective citizens of the Dominion. They were "an industrial asset," "ruder foreigners," or "manual labourers," expendable tools to enrich the Canadian nation and unworthy of the rights bestowed upon citizens. Sunera Thobani notes that citizenship status marks the divide between those who are considered "the legitimate heir(s) to the rights and entitlements proffered by the state"[43] and those outsiders who want to appropriate what does not belong to them. Certain immigrants are therefore excluded from the discourse of citizenship. In the past, exclusion was openly based on race, while today it is masked under the race-neutral language that targets seasonal workers and temporary migrant labour (although racialization of these categories still plays an important role in their construction).

Abiding by the principle that only working material should enter the country, on 7 April 1919, Parliament introduced Bill (No.52) to amend the Immigration Act and passed it on 12 May. The measure was in response to the escalating cost of taking care of undesirables, combined with the hardship created by unemployment conditions in the interwar period.[44] The late 1910s–early 1920s were also a time of great popularity for eugenic theories among Canadian scholars, scientists, and civil servants, and immigration of undesirables was a great concern to the eugenic movement.[45] Psychiatry and eugenics contributed to the belief that individuals with mental disabilities were the wrong kind of immigrants by reason of their natural tendency to behave in a dangerous and degenerate way.[46] Among other amendments, Bill (No.52) extended from three to five years the time during which immigrants could be deported. In order to dispel accusations that the move was in response to pressures exerted by the United States (where an identical provision was already in place), Mr Calder, minister of immigration and colonization, explained: "We are going to adopt that five-year period here, not because they have it in the United States, but simply because we think it is necessary that there should be a longer period than three years in which to ascertain whether or not many of these people who get into the country are desirable."[47] Another provision introduced in the bill was the inclusion within the prohibited classes of "persons who are suffering from chronic alcoholism," those showing "constitutional psychopathic inferiority," and "all persons who are either mentally or physically in such a condition as that they are not likely to be able to earn a living."[48] Asked to clarify the meaning of the term "constitutional psychopathic inferiority," Mr Calder replied, "In this classification are included various unstable individuals on the border line between sanity and insanity, such as moral imbeciles, pathological liars, many of the vagrants and cranks, and persons of abnormal sexual instincts."[49] Reiterating that these individuals represented a threat not only for the present but also for the future of the country, the Conservative member from Perth South (Ontario), Mr Michael Steele, warned the House that "a very large number of such people are in our hospitals for the insane throughout the country, a burden that is laid upon this country, that will continue for years to come and that will grow in cost to Canada, because we know that these feeble-minded people, if they are not confined in some institution, multiply very rapidly, and statistics show that about eighty per cent of their offspring are also feeble-minded."[50] For a country that was experiencing difficulties in taking care of its own growing population while simultaneously dealing with unemployment and the return of veterans, the fact that new additions could result in an economic burden was becoming the principal reason against the admittance of people who were likely to require medical care or services.

On 1 March 1928, during discussion on the budget, Mr James H. King (Liberal, Kootenay East, British Columbia) reminded his colleagues that, as representatives of the federal government, they had a constitutional responsibility towards the provinces, which were growing dissatisfied with the laxity shown by the government in permitting entrance to mentally insane persons. Mr King quoted the statements of officers in several provincial governments as examples of the dissatisfaction and financial difficulties experienced by the provinces. The testimony provided in 1922 by Dr Dunlop, "from the hospital for the insane, feeble minded and epileptic of the province of Ontario,"[51] summarizes the complaints of most provinces. Referring to the year 1922, Dr Dunlop observed:

> The proportion of those admitted born in Canada was 62 per cent. The proportion of the foreign born was 32 per cent. Many of the foreign born came from southern and eastern Europe. In 1922, we had 110 persons admitted who came from that part of the world and the cost of maintaining them was over $40,000. Very few of this class have any means, but become public paupers to be supported by the taxpayers of Ontario. One wonders, when dealing

with this subject, how long such an immigration policy should continue, and why the people of Ontario should be taxed to maintain such an alien element, who rarely become permanent assets but are simply floating liabilities.[52]

Echoes of these complaints are found in current discourses opposing the entrance of persons who for one reason or another are deemed likely to constitute a financial burden. As Thobani argues, the state uses citizenship as the status uniting deserving nationals, irrespective of their economic, gender, or racial characteristics, against undeserving outsiders who try to steal what is ours by right.[53] Far from exclusive to a few medical specialists, such beliefs were shared by several organizations, among them the Church of England Council for Social Service, the National Council of Women, and the National Child Welfare Association.[54] One of these organizations, the Social Service Council of Canada, passed a resolution at its meeting in Montreal calling for stricter application of the Immigration Act.[55] On 6 June 1928, Mr William F. Kay (Liberal, Brome-Missisquor, Quebec), presenter of the report of the Committee of Agriculture and Colonization, quoted the resolution in the House of Commons: "The Social Service Council of Canada having in mind the burden now laid on the various provinces of Canada in providing for the idiots and the insane, and the unfortunate results of admission into this country of the moron and feeble minded, desires to urge upon the federal government a continued and strict adherence to the provisions of the Immigration Act, in order that the door may not be open for the entrance of these types of prohibited undesirables."[56]

Mr King's intervention on the federal responsibilities regarding the provinces was occasioned during discussion around the McConachie case. In 1927, Mr McConachie, a Scottish immigrant resident in Canada, invited his wife and five children to join him in the new country; once in Halifax, after undergoing medical examination,[57] the parents were informed that the youngest of the children, a fourteen-month-old baby, had been diagnosed as mentally defective and was therefore inadmissible. Following the rejection, the child and mother were returned to Glasgow. As I will further discuss in the following chapter, the case received sustained attention in the press, generating feelings of sympathy among the public and contrasting opinions among politicians. Harsh criticism was reserved for the minister of immigration and colonization who, according to many parliamentarians, especially and unsurprisingly those sitting among the opposition, should have shown greater compassion providing the child with a ministerial permit to remain in Canada. On 5 March 1928, during discussion around the budget, Mr W.A. Boys (Conservative, Simcoe South, Ontario) took a chance for pleading the case with the minister: "If ever there was a case that merited the exercise of favourable discretion on the part of the Minister of Immigration ... it certainly was the case in question. I do not think for one moment that the minister would want to separate a baby of fourteen months old from its mother, nor, on the other hand, the mother and baby from the father and the other four little members of the family."[58] Four days later, Mr E.A. Peck (Conservative, Peterborough West, Ontario) called on the minister "to reverse his decision with regard to the McConachie child ... is there any harm in allowing that child to enter Canada when we get the advantage of receiving immigrants whom we need to the extent of a man, his wife, and four other children? There are weak-minded people in this country ... and it will not hurt very much if another weak-minded person is allowed to come in."[59] MPs were reluctant to face criticism from the press and several groups sympathetic to the family; Mr F.P. Quinn (Conservative, Halifax) noted: "The McConachie case has, up to the present time, elicited many expressions of sympathy and surprise. I am in receipt of letters every day from different people, and from various societies and organizations, expressing sympathy with the family affected."[60] Mr Quinn highlighted the discordance between the government's action and previous statements that Canada was "a land of tremendous possibilities and wonderful opportunities for the immigrant."[61] He pleaded with the minister to review the decision and admit the child.

Not everyone agreed. Several MPs considered it inappropriate to disregard the act, thus creating a dangerous precedent. Mr William R. Motherwell, minister of agriculture, asked his colleagues: "Why should these prohibited classes be allowed to enter the Dominion? ... If we are to allow the simple-minded to come in then what about the idiots, the imbeciles, the epileptics, and the others specified in section 3-fifteen or twenty prohibited classes? If one is allowed to enter, why not two; and if two, why not a thousand? ... We feel sorry for these unfortunates, but, nevertheless, we believe that, as in the past, they should be kept out. There can be no compromise on this question."[62] Mr J.J.E. Guerin (Liberal, St Ann, Quebec) also noted, "From what I can gather, this child is a cretin ... it was born an idiot and is bound to continue so ... I profoundly sympathize with the parents, but at the same time, dura lex sed lex. You cannot change the law."[63] Mr Guerin pointed out that it was one thing to be generous by spending money on an unproductive foreigner for a short period of time, but it was quite another to be burdened for years to come without receiving any benefit from it:

> If the expectation of life in the case of a child were brief it might be all right to admit it; but the fact is the children of this description live for

> many years. They grow up to the age of twenty or thirty, continuing in the same state of infantile decrepitude as when they were born ... such a child becomes a load on its parents ... To have a child of this sort in the house is therefore a misfortune beyond description ... I do not see how the Minister of Immigration, under the circumstances, could possibly think of admitting a person in that state.[64]

Ignoring feelings of affection and bonding between a child and her parents, Mr Guerin concluded his intervention by commenting, "It is not that I have anything but the deepest sympathy for the family, but it is a very easy matter for them to separate themselves from that poor unfortunate child. The baby would never be any the wiser, and the family would be much better off."[65] The opposite side did not show a better appreciation for human life; responding to those who defined the child as an economic burden, Mr A.E. Ross (Conservative, Kingston, Ontario) replied: "The medical officers say that in this type of infant not 9 per cent reach twenty to twenty-five years of age ... The baby has not one chance in a hundred to live ... If there was any possibility of a menace in this case, or the possibility of any liability being put upon us, I would say there was some reason in taking a little baby from its mother, but not in the case of a baby who has not got one chance in a hundred of reaching even twelve or fourteen years of age."[66] Encouraged by the support shown by several MPs, Mr Forke, minister of immigration and colonization, reported that "the child is to be left on the other side" while Mrs McConachie was rejoining her husband and four other children in Canada. The minister added, "I am led to believe that the child will not suffer from the loss of its mother; it would not understand the difference and no hardship will be caused the child by the loss of its mother."[67] His last remarks were addressed to the McConachie family, "The family has my sympathy. I hope Mr. McConachie and his little boys will become loyal Canadian citizens and forget the trouble they have gone through during the last year."[68] While documenting this case, I was surprised that the contentious infant was never referred to by her name, Margaret. The only reason her first name came to my attention is because it was reported in the *Globe and Mail.* For what concerned parliamentarians, Margaret McConachie was the baby, an idiot, a cretin or, at best, the McConachie child. She remained in the discussion just an impersonal presence. Naming is an act of love: we name our children, rename our lovers with terms of endearment, find names for our creations. We leave nameless what we do not care about. The omission of Margaret's name from the public record is a further insult I refuse to conceal.

The McConachie case reached a satisfactory conclusion (from a parliamentary perspective as it is unlikely the McConachies shared the same satisfaction); still the issue of admitting immigrants who, due to mental or physical disabilities, were likely to become a burden for taxpayers remained on the table. Noteworthy was the pressing situation of provincial governments and municipalities that felt the burden directly on their shoulders, particularly in the context of economic hardship created by the Great Depression. On 10 July 1931, Mr Wesley A. Gordon (Conservative, Timiskaming South, Ontario), new minister of immigration, emphasized: "In the poorer municipalities ... the burden of taking care of people who are not able to take care of themselves, by reason of physical or mental defects, is thrust upon the public. And those municipalities are having a hard enough struggle as it is to get along, without having that additional burden thrown upon them."[69] In light of these concerns, deportations continued. Between the years 1930–1 and 1944–5, 2,724 individuals were deported from Canada for medical reasons, a majority of them (1,596 or 58 per cent) for mental diseases.[70] The number was a significant reduction from previous years: gripped by the Depression, the decade leading up to the Second World War witnessed a decline in both the international movement of people and efforts to find out and deport those who entered the country.[71]

Given that the problem was financial in nature, what of immigrants who were able to support themselves? The 1910 act eliminated the option for mentally deficient individuals to enter Canada when accompanied by a family member willing to support them, but the question remained whether it was possible to make exceptions if advantageous to the receiving country. One of such cases came to the attention of Parliament in 1938 when an American citizen, Mr Harper, asked for his sister to be accepted into Canada. An American-born woman and the widow of an Italian citizen, Mrs Belle Hervey Harper Cazzani had developed "mental trouble" and her brother wanted her close by so to take adequate care of her. The case would have been largely ignored in the Senate if not for the £55,000 the lady would bring to the country. On 4 April 1938, Hon Mr Lacasse (Liberal, Ontario) moved second reading of Bill M-1, an Act respecting Madame Belle Hervey Cazzani. Beyond the outcome of the case, my main interest is in examining the role disability played within the Immigration Act vis-à-vis its financial implications. Opinions tended to differ among senators: while some concurred with Hon Joseph Philippe Baby Casgrain (Liberal, Quebec) that "Canada would benefit by the money that this woman would bring in, and when she died her estate would pay succession duties to the Province of Ontario,"[72] others believed that the negative consequences of accepting Mrs Cazzani would outweigh the benefits. Quoting a statement by Mr Frederick Blair, director of the immigration branch, Hon Raoul Dandurand (Liberal, Quebec) warned: "The present Bill, so far as we are aware, is the first effort to accomplish the admission of a prohibited person by means of an Act of Parliament. Should it be successful it will establish a precedent that will quickly be seized upon by many others desiring similar concessions."[73] Such a precedent could have opened the door to an unforeseeable number of mentally disabled immigrants, who could or could not have the means to cover their expenses. Should the immigrant be unable to cover said costs, the burden would fall to local governments. The potential economic obligation was a threat far too big to ignore. Economic considerations had at this point taken over the debate around inadmissible immigrants.

Phase Two: Late 1940s to Early 1980s

Whereas the late 1940s were years of economic prosperity and low unemployment, conditions associated with an open-door immigration policy, selectivity remained a priority.[74] Persons with diseases or disabilities were regarded as unemployable. In 1953, Mr J.S.A. Sinnott (Liberal, Springfield, Manitoba) reminded the House that "a couple ... slipped through the immigration department a year or two ago when suffering from a serious malady ... That couple is now costing the municipality $40 a month or $480 a year ... That is the type of immigrant we should keep away from this country."[75] It was one thing to help disabled persons born within Canada, but it was a whole different matter to take care of those who, being born outside of the country, had no rights to assistance. Still, while uneasiness with inadmissible immigrants continued unabated, the reasoning behind it had by now switched from a concern with newcomers likely to corrupt the country's genetic pool to more mundane economic considerations. According to Robert Menzies, the change was solidified by the war against Nazism: the growing association between Nazi theories and eugenics had forced many Canadians to distance themselves from hereditary theories, scientific and biological racism, and the associated emphasis on the risks of degenerate populations.[76] Furthermore, those years witnessed an increasingly strong demand for Medicare throughout Canada;[77] in the event this movement succeeded, the last thing on Ottawa's agenda was to get burdened with the medical cost of immigrants.

The late 1940s and the 1950s were years of moderate but steady advancement for disabled citizens, partly due to the impact of returning veterans. Whereas disabled veterans had employed strategies such as organizing and mutual support following the First World War, it was only after the Second World War that these efforts gained prominence.[78] The Canadian government made the return of veterans, including disabled veterans, to civilian life, one of its priorities.[79] The Department of Veterans Affairs was created in 1944 and provided disabled veterans with free support and medical services at a time when approximately one-third of Canadians did not have private medical insurance, not to mention there was a huge effort to reintegrate them into the workforce and civil society.[80] The effectiveness of the department in helping veterans readjust to civil society was the result of two factors: first, Canada wanted to avoid repeating the mistakes incurred after the Great Conflict when no serious effort had been made to reintegrate and financially support veterans; second, it was the outcome of a new determination among Canadians that a planned economy and expansion of government social welfare implemented in wartimes should continue in peacetime. The department played a major role in the maintenance and expansion of the welfare system, and it represented a model for the Hospital Insurance Act passed in 1957 and for Medicare in 1968.[81]

Within society at large, new organizations advocating on behalf of persons with disabilities were born, such as the March of Dimes Canada, the National Institute for Mental Retardation, the Canadian Paraplegic Association, and the Canadian Association for the Mentally Retarded. Legislation was eventually passed in Parliament: among others, Bill 462 to provide for allowances for disabled persons, passed on 8 June 1954; the Blind Persons Act Amendment Lowering Eligible Age to 18 Years and Increasing Maximum Annual Incomes, received royal assent on 28 June 1955; the Disabled Persons Act Amendment to increase payments and allowable income, received royal assent on 15 February 1962, together with a similar amendment to the Blind Persons Act; and Bill No C-125, An Act to Amend the Old Age Assistance Act, the Disabled Persons Act, and the Blind Persons Act, increased the monthly amount of money available for assistance or allowance and received royal assent on 12 December 1963. The trend was towards recognition of Canadians with disabilities, from the initial tentative efforts in 1952, when Mr Arthur Masse (independent Liberal, Kamouraska, Quebec) spoke of the duty of the government "to help the handicapped" provided it was not "detrimental to our economy"[82] to the statement made two years later by Mr Vaillancourt (Liberal, Quebec) who, in congratulating the prime minister on the introduction of Bill 462, emphasized how the support provided by the government was going to transform disabled persons from burdensome to useful members of society.[83] On 15 June 1956, in the House of Commons Mr G.W. McLeod (Social Credit, Okanagan-Revelstoke, British Columbia) talked about the contribution disabled citizens could make to Canadian society:

"we are beginning to find that men and women who a few short years ago we thought could not be used in any manner in our labour force are now being trained so they can assume a fitting place in the circles of labour, provide a livelihood for themselves and contribute to the economy of Canada."[84] Mr McLeod lauded the new program in technical training for people with special needs made possible under a joint provincial and federal financing agreement.[85] While still trapped in understandings of individuals as worthy insofar as economically productive, the statements were a moderate improvement for a House of Commons that had previously shunned disabled persons.

This new orientation built momentum in subsequent decades. The year 1964 witnessed the first Canadian federal-provincial conference on mental retardation, which followed in the steps of a 1952 federal-provincial conference on disability. The federal government organized both conferences in response to pressures from social movements headed by disabled citizens (veterans in particular), parents of children with disabilities, and professionals. Whereas the goal of the 1952 conference had been to expand employment and training programs for disabled adults across Canada, the purpose of the 1964 conference was to improve health, welfare, and educational and vocational services for mentally retarded children and adults. Among the achievements registered by the end of the conference, there was an expansion of resources and opportunities in the community made possible by a special funding program whose costs were shared between the federal and provincial governments (the Canada Assistance Plan).[86] On 19 February 1973, during discussions surrounding a motion concerning measures to aid handicapped persons, Mr W.G. Dinsdale (Progressive Conservative, Brandon, Manitoba) emphasized that "people with physical handicaps, or for that matter with any disability, actually have a special quality that places them in a position to make a particularly effective contribution to the society in which they live ... disability is far more frequently the result of social, emotional or economic dependency than physical impairment."[87] He invited his colleagues to work for "a society where there is a genuine respect for the handicapped ... where the handicapped have a fundamental right to participate in industry and in society according to their abilities; where socially preventable distress is unknown; and where no man has cause to feel ill at ease because of disability."[88] Unfortunately, things were not looking as rosy for those seeking admission to Canada and the provision of medical admissibility stood. As the post-war years saw the implementation of measures concerning health and welfare,[89] the government was determined to prevent the entrance of individuals who could take advantage of these new services at the expense of Canadian citizens.

As Dossa and Thobani have argued, the welfare state did actually represent a setback for immigrants insofar as they were increasingly pictured as outsiders threatening to deplete resources at the expense of citizens.[90] Pressured by various organizations dedicated to improving the lives of Canadians with disabilities, Parliament passed more inclusive legislation; similar pressures were, however, absent on behalf of foreign disabled individuals.

In the post-war era, Canadian society registered a new interest in human rights, resulting in a more accepting attitude towards ethnic diversity and the growth of a human rights movement in Canada.[91] Yet, when it came to immigrants, disability issues were never addressed. The Immigration Act of 1952 continued to apply the principle of exclusion by listing all prohibited classes and establishing the requirement of medical examination for immigrants.[92] Even the removal of racial barriers in the following decade did nothing to eliminate discrimination on the basis of disability. Ethnicity was replaced by an emphasis on skills, but immigration remained a tool to benefit the national economy,[93] something disabled persons were considered unable to do. This was in line with the emergence of neo-liberalism and its celebration of efficiency and the market economy. According to Raymond Plant, neo-liberalism has been presented by its supporters as compatible with principles of equality of opportunity and anti-discrimination under the argument that, within the market economy, workers are assessed based on efficiency rather than on grounds of gender, race, and religion.[94] The neo-liberal argument of racial-blindness has been debunked in the analysis of scholars who have clearly demonstrated how "race and racism are inextricably embedded" in the neo-liberal project insofar as immigrants are exclusively assessed based on their productivity and contribution to the state's growth, whereas citizens are not subject to such stringent rules and requirements.[95] Furthermore, the analysis of neo-liberalism's effects on the physical and social body is incomplete if we ignore the discrimination it perpetuates against persons who, due to mental or physical disability, are perceived as inefficient and therefore not useful to the market economy. I agree with Dossa that neo-liberal discourse has certainly not "advanced the interests of people who do not fit the criteria of young, able-bodies males, perceived as productive members of society."[96] If anything, neo-liberal discourse rests on the assumption that discrimination against those individuals is unproblematic.

Despite the absence of substantial modifications to the admissibility provision, minor changes were implemented with respect to the criteria of rejection. In 1968, Bill C-30 was introduced in the House with the purpose of giving permanent status to immigrants following recovery from mental disorder.[97] As Hon Lionel Choquette (Progressive Conservative, Ontario) noted in the Senate, "Unfortunately, in our present society, mental illness seems to attach a stigma which remains with a person for the rest of his life."[98] Within this context, the bill was a refreshing improvement. Also deserving consideration is the fact that, during discussion around Bill C-30, several MPs felt discomfort at the language used in the Immigration Act to define persons with mental disabilities. The issue had already come up two years earlier when, with reference to section 5 of the Immigration Act, Mr J.A. Irvine (Progressive Conservative, London, Ontario) had pointed out that "this act ... is now out of date ... in the very language it employs. The words 'idiot', 'imbecile' and 'moron' have not been found in modern medical literature for many years. They are considered repulsive as well as outdated. The same can be said of the word 'insane'; and medicine has yet to define the word 'psychopathic.'"[99] In 1968 a suggestion was made to amend the language of the act. On 22 March 1968, the House discussed Bill C-30, to amend the Immigration Act, giving permanent status to immigrants who had suffered from a mental disease but recovered. While dealing with the section concerning immigrants with mental disabilities, Mr John Gilbert (NDP, Broadview, Ontario) remarked: "The word 'insane' is not generally accepted by Canadians today. It connotes something that we do not really like to think about. We have got rid of the title 'insane institutions' and call them mental institutions. I would hope there would be a further amendment to the section and the word 'insane' changed to the phrase 'mental illness.'"[100]

Once established that people who recovered from mental illness should be allowed into Canada, the next step was extending the relaxation to persons who, though still "mentally retarded" and a potential economic burden, had members of their family willing to cover any expenses. In November 1968, the House of Commons began discussing Bill C-10, to amend the Immigration Act, allowing for conditional admission of mentally retarded persons. As Mr Hubert Badanai (Liberal, Fort William, Ontario) observed, "if the person seeking admission is a member of a family already in Canada, financially responsible and capable of giving satisfactory security against such an immigrant becoming a public charge, in such a case there would appear to be sufficient justification to relax the regulations to enable that person to join his family."[101] The government responded that the bill was unnecessary given the minister's discretionary powers to act through individual permits. Mr Gerard Loiselle (independent Liberal, St Anne, Quebec), parliamentary secretary to the minister of manpower and immigration, explained that the previous and current ministers had already "relaxed the absolute prohibition on the admission of those afflicted with mental retardation ... and allowed them to enter under ministers' permits."[102] The Immigration Act was, therefore, in no need of further amendments and indeed no change of the kind would go through until the passage of the IRPA in 2001. While the attempt to modify s. 5 of the Immigration Act failed in 1968, the provision continued to be seen as problematic both in its language and content. In 1975, the third report of the Special Joint Committee of the Senate and House of Commons on Immigration Policy dealt with its most evident shortcomings. Basing its findings on input received by Canadians who participated in public hearings held across the country, the report concluded that contrary to what stated in s. 5, "immediate members of a family should not be separated because one member suffers from mental retardation."[103] The committee "recommends that sponsored dependents who are mentally retarded be admissible."[104] It also noted: "Because many forms of mental illness and epilepsy can now successfully be treated and controlled, most Committee members agree that a person with a history of such a disease should be admissible providing he can lead a normal and useful life. A minority of the Committee would have eliminated mental illness and epilepsy altogether from the prohibited classes."[105]

If previous years had seen some discussion but little achievement, big hopes were cast in 1976 on what is traditionally portrayed as a milestone in Canada towards a non-discriminatory immigration policy. Bill C-24, the Immigration Act, 1976, marked the third time in that century, after 1910 and 1952, for a complete overhaul of immigration legislation. The bill was the result of four years of preparation started in 1973, when the minister of manpower and immigration created a group to work on a Green Paper intended to be a vehicle for public debate. The Green Paper was tabled in Parliament in February 1975 and referred to a special joint committee of the Senate and House of Commons. Recommendations made by the committee formed the basis on which Bill C-24 was formulated.[106] For persons with mental or physical disabilities seeking admission to Canada, the new Immigration Act could have represented a dramatic change towards a less discriminatory policy. Expectations went unmet, however: the language was different and gone were references to idiot or lunatic, but the substance was unaltered and disabled persons remained unwanted. This reality was quite disappointing when considering that since the 1950s a number of advancements had been made on issues of human rights as well as anti-discrimination practices. Mr E.W. Woolliams (Progressive Conservative, Bow River, Alberta) remarked: "The bill requires health examination of persons outside Canada to determine whether they will be admitted to Canada. I thought the human rights bill protected those who have some physical or other disability to enable them to function in society like the rest of us, without discrimination. This bill totally contradicts the human rights bill."[107] Additionally, the bill was formulated in general terms with the understanding that more precise rules would be established by way of regulations unavailable for discussion in Parliament. Several MPs perceived the practice as a rebuff of their prerogative to assess, debate, and pass legislation.

On 14 March 1977, while the House of Commons was discussing a number of proposed amendments to Bill C-24, Mr J.R. Holmes (Progressive Conservative, Lambton-Kent, Ontario) commented, "the bill does not contain most of the important provisions of the government's immigration policy. It is readily apparent that a good deal of the policy will be established by way of regulation and order in council ... I find it inexcusable that the major, operative portion of legislation, which is embodied in regulations, is not available for members of the House of Commons to assess."[108] The practice was far from inconsequential. Establishing if a person's admittance "would cause or reasonably be expected to cause excessive demands on health or social services"[109] was ultimately left in the hands of medical officers. Mr Holmes felt "somewhat uneasy that a medical officer would be given unlimited discretionary powers to make such a decision."[110] He remarked, "we have a different perception of a medical problem today as compared to 20, 30 or 40 years ago" and "certain medical problems which may be considered inadmissible today, as epilepsy was in 1952, may appear archaic in the future with the advancement of medical technology."[111] He added: "It is not inconceivable that a medical officer in the future could deny admission to an immigrant on medical grounds and yet another medical officer could permit another individual with the same medical problem to enter Canada. This ... does point out the difficulties of giving unlimited discretionary powers to a medical officer during an era where there are rapid changes in medical technology."[112] Still today, the IRPA considers as inadmissible those who, according to a medical practitioner appointed by the state, are deemed likely to constitute an excessive burden on health and social services. The discretionary powers given the medical practitioner[113] remain problematic; in many instances, it is a guess for any doctor to assess if a medical condition that is under control will become significant in a period of time from five to ten years. Equally concerning is the absurdity of accurately guessing what will occur in the future when, today more than ever, predictability in the medical and scientific fields has become a lost game. And yet, as noticed in the previous chapter, we continue to assume that science and medicine are infallible. They have been given complete authority to describe our bodies, and that description goes unquestioned.

Phase Three: 1980s to 2002

The 1980s represent a unique case for the issue at the core of this book. The provision of medical admissibility received little attention in Parliament as the 1981–2 recession created an anti-immigrant backlash that resulted in declining numbers of immigrants accepted. Something else of importance occurred, though: the Canadian Charter of Rights and Freedoms came into being as part of the Constitution Act of 1982. It replaced the Canadian Bill of Rights passed in 1960 by the Diefenbaker government; whereas the latter applied only to federal legislation, the Charter is part of the constitution and applies to federal and provincial governments, empowering the courts to review and invalidate any law contrary to it. The Charter was reflective of a new attitude towards equality and rights that went beyond the mere legal sphere to touch upon every aspect of society. I will discuss in more depth its effects within the larger society later on; for now, let us focus on the way this new impulse towards equality manifested itself in parliamentary discussions surrounding immigration.

The debates in the House of Commons reveal that the attention to rights that was developing within society quickly penetrated the thick walls of the Parliament buildings in Ottawa. In March of 1985 the House of Commons debated Bill C-27, An Act to Amend Certain Acts Having Regard to the Canadian Charter of Rights and Freedoms. In his intervention, Mr Neil Young (NDP, Beaches, Ontario) complained about the scarce opportunities given to persons with disabilities within Canadian society. On the issue of immigrant applicants with disabilities, he argued: "Our Government could set an example for other governments in the world by asking if it is fair to continue to deny a family the right to reunite simply because a child whom they were forced to leave in the country from which they emigrated happens to be developmentally handicapped – commonly referred to as mentally retarded – and the child would be a burden on the state in terms of health costs."[114] In the late 1980s, for the first time, acting on behalf of their constituencies, several MPs brought petitions to the attention of the House asking for a revision of the medical standards for immigrants. The petitions presented by Ms Barbara Greene (Progressive Conservative, Don Valley North, Ontario) on 8 June 1989, and by Mr Mac Harb (Liberal, Ottawa Centre) on 15 December 1989, justified the need for a revision in terms of "compliance with the Canadian Charter of Rights and Freedoms and the enhancement of the human rights and dignity of persons with disability."[115] These petitions signalled a new awareness among members of the disabled community with respect to the potential of the Charter in enhancing their rights. With respect to medically inadmissible persons, the Charter represents the most serious challenge to a provision that keeps discriminating against them. This point has been true since 1985 when the Supreme Court – with the Singh decision – established that everyone physically in Canada, non-citizens included, is entitled to the full protection of the Charter.[116]

Following passage of the Charter, there was an increase in the number of parliamentary interventions that contested the rejection of certain categories of immigrants on the grounds that it violated section 15(1) of the Charter. On 19 June 1992, while discussing Canadian immigration policy, Mr Warren Allmand (Liberal, Notre-Dame-de-Grace, Quebec) reminded his colleagues that one of the Charter's objectives was "to ensure that any person who seeks admission to Canada is not discriminated against in a manner which is inconsistent with our Charter of Rights."[117] The year 1992 witnessed intense debate on immigration and particularly the clause of medical admissibility. On 8 June 1992, during discussion around Bill C-78, An Act to Amend Certain Acts with Respect to Persons with Disabilities, Mr Rod Murphy (NDP, Churchill, Manitoba) asked for repeal of s. 19(1) of the Immigration Act, which "prevents most persons with disabilities from getting landed immigrant status and from entering Canada."[118] Mr Murphy explained, "this provision alone serves to perpetuate the misconceptions and misunderstandings about the abilities of disabled persons."[119] He received the endorsement of other MPs; Mr Ronald J. Duhamel (Liberal, St Boniface, Manitoba) recognized that "not all people with handicaps are a burden to the state. Quite the contrary, they can and often do make major contributions."[120] Mrs Beryl Gaffney (Liberal, Nepean, Ontario) condemned the act for failing to acknowledge that "many persons with disabilities are able to be productive members of society and that they are eager to make a meaningful contribution to the Canadian economy."[121] The fact that these issues were discussed in Parliament in a decade marked by the desire for fiscal conservatism and restraint in government spending speaks volumes about the impact that the Charter as well as a growing advocacy movement were having on Canadian society.

Analogous sentiments emerged during debate around Bill C-86, a Measure to Amend the Immigration Act. The bill touched on different aspects of immigration, from the proposed requirement that immigrants settle in certain regions to changes to the health provision of admissibility. In his intervention, Mr Warren Allmand reiterated the importance of eliminating from the act any provision that could discriminate among people "in a manner which is inconsistent with our Charter of Rights."[122] In the Senate, the medical admissibility clause also came under attack by senators who saw it as violating the principles embodied in the Charter. Senator Mark Lorne Bonnell (Liberal, Prince Edward Island) warned that "even with the amendments to section 19(1) which remove references to 'disease[s]', 'disorders' and 'disability', the term 'excessive demand' in reference to health and social services, could exclude persons with disabilities. This is intolerable ... The international community considers Canada to be a caring nation. It is time our actions fulfilled our commitments to the protection of human rights."[123]

Bill C-86 was then brought into committee where the opinions of experts were canvassed and discussed. Since the early stages, the debate was heated. On 27 July 1992, Mr Allmand pointed out that before further consideration, the act should clearly state whether "the purpose of the changes is to deal with health problems that you're trying to correct or with financial problems in respect to our health and social services."[124] Under the act, medical officers were given authority to assess whether the admission of an individual represented an excessive demand for health and social services. Mr Allmand wondered if the provision made sense at all since "the medical officer is trained to deal with health problems and not financial problems,"[125] hence, "how can you ask a medical officer to deal with questions of financial services and the impact on the services of the country?"[126] No answer was recorded. The attention then shifted to the fact that Bill C-86 could represent a major change in Canadian immigration by adopting a fairer approach towards persons with medical illnesses or disabilities. Mr Brian Grant, director of the control policy within the Department of Employment and Immigration, noted that "when we speak about control and screening people as they come into the country, we are essentially no different from any other sovereign state."[127] He added, "We have removed the reference to disability from the act to address the perception that the act discriminates against a group of people or individuals. We have also removed the reference to disease, disorder or health impairment."[128] He rejected accusations that the definition of excessive demand was too vague: "The definition itself ... is in regulation 22 ... It speaks of the availability of services, of the accessibility of services, whether Canadians are lined up, whether they would be displaced if somebody were brought in requiring that service, and the cost of that service. We propose to develop in regulations a list of services that are either not available or that are in critical short supply in the country ... We are also looking at developing a factor to deal with cost."[129]

After the government presentation, experts from various organizations were called in. Mr Jim Derksen, president of the Canadian Disability Rights Council, a group organized by the Coalition of Provincial

Organizations of the Handicapped,[130] declared his dissatisfaction with the bill. According to Mr Derksen: "The actual impediments in the act that discriminate against people with disabilities have antecedents that go back to ... 1869 ... The actual functioning of that Immigration Act of 1869 has not really changed with the various immigration acts under which we live today, and, we would submit, with the amendments being proposed. The assumption still is that people with disabilities will be an undue burden on society."[131] Ms Yvonne Peters, executive director of the Canadian Disability Rights Council, agreed that "first and foremost, section 19 of the act ... conveys the message that the participation of people with disabilities in our society is not welcomed, is not valued ... from a symbolic sense, disabled Canadians perceive that they are less valuable than other citizens in Canadian society."[132] Furthermore: "Canada holds itself out as a leader in promoting human rights in the international arena. Canada ... is a signator to a host of international agreements and instruments that profess its commitment to uphold and respect various human rights principles ... they include the Universal Declaration of Human Rights and Canada's willingness to declare the Decade of the Disabled, which was promoted by the United Nations, and so on. It is therefore ironic that Canada would continue to give credence to a law that is so patently offensive and invidious to people with disabilities."[133] She dismissed as inconsequential the omission in the bill of any "explicit reference to disability."[134] The removal represented an improvement from previous legislation insofar as it "eliminates evidence of overt discrimination,"[135] yet "removing the word disability does little to reassure people with disabilities that the section ... will not be used to continue to exclude people with disabilities from immigrating."[136] With or without mention of disability, "the admissibility of people with disabilities into Canada is couched in what we believe are negative terms of danger to public health and safety and excessive demand on health and social services."[137] Historically, different versions of the Immigration Act had excluded persons with disabilities. The proposed amendment was no different and relied on "archaic laws introduced at a time when people with disabilities were systematically isolated from mainstream society and regarded as social outcasts."[138] Equally troubling was the decision of having two doctors assess whether the perspective immigrant was admissible. Considering that doctors had repeatedly counselled women carrying disabled fetuses to abort and parents of disabled children to place them in institutions, it seemed questionable that the medical profession was given carte blanche in deciding "who is a worthy candidate to get into Canada and who is not."[139] The Canadian Disability Rights Council suggested the alternative of "a three-person committee that would review the physicians' decisions and would ensure that human rights and equality rights were upheld and respected."[140]

The following speaker was Ms Diane Richler, executive director for the Canadian Association for Community Living, a Canada-wide charitable organization of family members and others working to improve the lives of persons with mental disabilities. Ms Richler worried about the image Canada was presenting to the international community: "For many future Canadians the first experience they have of what Canada is like as a country ... is what they experience when they apply to Canada as immigrants. Unfortunately, for families who have a member who has a disability, the messages that are sent ... are very, very negative ... People are made to feel that being disabled or being related to a person with a disability is a crime in Canada."[141] Mr Jerome Di Giovanni, secretary of the Canadian Disability Rights Council, was next. Mr Di Giovanni also represented the Quebec Multi-Ethnic Association for the Integration of Handicapped People/Association multi-ethnique pour l'intégration des personnes handicappée du Quebec, a non-profit organization of persons with disabilities and parents of persons with disabilities, created with the goal of fighting the provision of medical admissibility. Branding section 19(1)(a) "offensive and discriminatory," Mr Di Giovanni demanded its withdrawal. He dismissed the concept of excessive burden as "a paranoid concept. No government, be it Conservative or Liberal, has ever managed to demonstrate that disabled persons were an excessive burden on Canadian society."[142] As stated in the brief to the proposed amendments submitted by the Canadian Disability Rights Council: "There is no data on the rate of use of 'health and social' services by immigrants generally as compared to the Canadian population at large – let alone on the rate of usage by immigrants with disabilities. Until a comprehensive study is conducted on this subject, 'excessive demand' should not be used as the statutory test for exclusion."[143] The brief argued: "The assumption of 'excessive demand' is systematically used to exclude persons with disabilities while applicants of high health risk categories such as smokers, heavy drinkers and high-stress workaholics are not assessed on the same basis. So 'cost arguments' are being used selectively to exclude only certain categories of applicants – among them, persons with disabilities."[144] Di Giovanni referred to the example of Mr Maurice Lwambwa Tshany, who was fighting a removal order from Canada:

> Maurice Lwambwa Tshany is an artist. He has opened his own workshop in Montreal. You can see how this makes him an excessive burden for Canadian society. Maurice Lwambwa Tshany teaches pottery and African art to young Canadian men and women in Montreal. That certainly makes him an

> excessive burden on Canadian society. There were exhibitions of Maurice Lwambwa Tshany's works in Vancouver ... in the Maritimes, in Montreal and in Europe. That surely makes him an excessive burden for Canadian society. Maurice Lwambwa Tshany sits on the board of the Multi-ethnic Association for the Handicapped People. That surely makes him an excessive burden ... Maurice Lwambwa Tshany decided to take an active part in the cultural, social and economic development of our society. And yet, Maurice Lwambwa Tshany will soon be expelled from Canada. Why? Because he has a physical disability, because he is in a wheelchair.[145]

While the decision to focus on the experience of Mr Lwambwa Tshany, a man of obvious resourcefulness and creativity, was tactically effective, it is inherently unfair to demand exceptional behaviour from persons with disabilities as a precondition for recognizing their worth. Exceptionality is not demanded of the able-bodied, and it should not be demanded of the disabled.

The last speaker for the day, Mr Gerry MacDonald, vice-chair of the Coalition of Provincial Organizations of the Handicapped, explained that although "Canada has the right of sovereignty and can decide who will and who can cross its borders,"[146] he saw as problematic the fact that the section under scrutiny patently contravened section 15(1) of the Charter. According to paragraph 3(f) of the Immigration Act, "Any person who seeks admission to Canada on either a permanent or temporary basis is subject to standards of admission that do not discriminate in any manner inconsistent with the Canadian Charter of Rights and Freedoms."[147] As indicated in the brief to the proposed amendments, the paragraph meant "there can be no discrimination against immigration applicants with disabilities (and refugees) at any point in the application process."[148] Another element of contention was the decision to rely on two medical practitioners when assessing immigrants with disabilities: "How do you judge the future potential ability of a person with a disability to be independent and self-sufficient through a medical examination?"[149] Before opening the floor to questions from the committee, the last word was left to Mr Derksen who concluded with an anecdote. Speaking to his experience as an advocate for disability issues in Parliament over the previous decade, he remembered: "I met Paul Martin, Sr. once and we had a great chat. I found out that he had a disability similar to my own – polio, once very common but now not common among younger people in Canada. I should say I don't think Paul Martin or I have ever been an undue burden on social services or medical systems in this country."[150] Reactions to the statement are not recorded, but it is a likely guess it was met with general consensus.

Mr Derksen's statement invites a reflection on the reasons behind society's definition of immigrants as outsiders to be assessed based on a whole different set of criteria. In the mind of most Canadian citizens, Paul Martin Sr was never a burden, but is the same true of a foreigner? Immigrants remain "strangers within our gates."[151] The state continues to legitimize a discourse that relies on the distinction between citizens and outsiders, with the latter "depicted as making unreasonable claims upon the nation and its precious finite resources."[152] Citizens enjoy privileges outsiders cannot ask for. In this sense, even the unattractive label of disabled is something that non-citizens can only dream of as they will never be able to claim it for themselves here in Canada.

After meeting representatives of social organizations, the committee heard from government officials working in sectors related to the administration of the Immigration Act. They were called in to clarify the process for assessing immigrants and to provide a rationale for acceptance of the proposed amendments. Dr Neil Heywood, assistant director at the Immigration and Overseas Health Services within the Department of Health and Welfare, reiterated that the removal of the words "disease, disorder, disability or other health impairment" eliminated the "perception of inequality" and "potentially offensive terminology while leaving the effective elements of the medical assessment process."[153] His statement confirmed what previous speakers had identified as a camouflage allowing the state to continue discriminating against persons with disabilities by using a more nuanced language. Words were changed, but the result was the same. Dr Heywood also attempted to disperse the perception that the act operated to automatically exclude persons with disabilities from immigrating. He explained: "Each applicant is assessed individually ... An applicant may be determined to be unlikely to create excessive demands on health or social services and be capable of employment and self-support, in which case she or he will be medically admissible."[154] This observation validated Ms Peters' and Mr MacDonald's concerns regarding the fact that two medical practitioners were given sole authority to decide upon something they had no expertise in, namely if the individual was employable and self-supporting. Dr Gilles Fortin, acting director at the Immigration and Overseas Health Services within Health and Welfare Canada, responded to these concerns by explaining that the department merely provided advice to the immigration department. The latter had the final word and, therefore, the responsibility to take into account other factors such as "social consideration, family considerations, economic considerations."[155] However, when pressed if Department of Health and Welfare physicians had the power to stop someone from coming into the country, he conceded that was the case.[156] Although the immigration department had the authority of "by-passing these medical restrictions by providing the applicant with a minister's permit,"[157] Dr Fortin's statement confirmed the fears expressed by the Canadian Disability Rights Council in the brief to the proposed amendments. In the document, the department was accused of abusing the issuance of permits and the practice was subject to severe criticism: "While ostensibly they [ministerial permits] are used to alleviate hardship in individual situations, it must be remembered that the hardship is the result of a discriminatory practice in the first place. Secondly, their use perpetuates systemic discrimination against immigration applicants with disabilities by serving as the main weapon in a strategy to 'challenge-proof' an unconstitutional law ... Thirdly, the current use of Minister's Permits, reinforces the idea that persons with disabilities are exceptions – requiring 'special' not equal treatment."[158] Indeed, ministerial permits represented the exception, whereas the medical assessment remained the measuring stick in deciding admissibility.

Additional external experts were then called in. Dr Nicholas Birkett, associate professor in the Department of Epidemiology and Community Medicine at the University of Ottawa and member of the Canadian Public Health Association, addressed the issue of excessive demand. Against fears that "immigrants are going to deprive Canadians of their health and social benefits," Dr Birkett stated: "There is no evidence that this is the case. On the contrary, data from a number of studies suggested that immigrants actually make lower demands on health and social services than do non-immigrants. In many cases this occurs because immigrants make use of lower-cost alternatives and volunteers."[159] He recognized that the system was under pressure but refused to put the blame on immigrants, instead identifying the main causes of the crisis as coming "from our aging population, from the insistence of both the general population and physicians on the need for high-tech medicine, on heroic attempts to save dying patients, and on the high unemployment rates."[160] Dr Pran Manga, chairman of the Department of Health Care Administration at the University of Ottawa, explained: "There are, indeed, a good number of multicultural health studies in Canada that show immigrants do not visit doctors as often as they should. They do not use a whole variety of other health services as frequently and to the same extent as non-immigrants, meaning Canadians."[161] This discrepancy resulted from a number of factors: "The extended family idea is one. They look after themselves more ... They have different concepts of what it means to be ill or not well. There are linguistic barriers. There are cultural barriers. There are barriers of economics, meaning the poor immigrants are not likely to go to doctors who are far away."[162] Immigrants' health cost was part of a bigger picture: "How do we assess offsetting benefits? Immigrants pay taxes when they enter Canada. They also contribute to the general economy and the status of Canada around the world. How do we determine if the investment of up-front medical costs is balanced by the future potential of an individual?"[163] There were further questions. Was excessive demand going to be measured in terms of dollars or looking at the availability of health services? If the latter was the case: "How widely do we set our net? One example would be that there have been some predictions ... of a serious shortfall in the availability of cancer treatment facilities in Ontario. An immigrant who has smoked heavily for 20 years would have a high risk of developing cancer. Do we exclude this person from immigration on the grounds that there is a reasonable expectation that they would be demanding a limited resource?"[164] Also important was the potential unconstitutionality of the provision excluding immigrants with disabilities as "to deny people immigration solely on the basis of a disability would, in our opinion, be a violation of the Charter of Rights and Freedoms, which this act is subject to."[165]

Dr Birkett was then confronted with the question of how to balance these concerns with the possibility of having to spend considerable amounts of money caring for disabled or diseased immigrants. A member of the committee, Mr D.J.M. Heap (NDP, Trinity–Spadina, Ontario), recalled a "few horror stories"[166] that had occurred in Canada. The doctor responded that those stories represented mere "anecdotal cases" and that a general policy "should be based on the totality of the system rather than on one individual case."[167] He drew on the example of the existing welfare system: "It is like saying we should not have welfare because a case can be found of somebody who has committed welfare fraud. That's wrong; he shouldn't commit welfare fraud, obviously, but that doesn't mean the system is inappropriate and should be changed."[168] Dr Birkett noted that the number of people with disabilities trying to enter Canada was limited as were the costs: There are 1989 statistics that were released by Employment and Immigration. Only ... about 0.25% of the applicants, were deemed to be medically permanently inadmissible under the criteria in existence at that time ... We're not talking about the system screening out large numbers of people who would flood into Canada."[169] In addition, "health is only one aspect of the requirements to immigrate to Canada, and somebody who is in fact at death's door, who is going to require major costs, would be unlikely to meet the other criteria that would be set unless there are humanitarian reasons or family reunification or some such arrangement, in which case the humanitarian considerations might outweigh the costs."[170] Dr Birkett believed that immigrants with disabilities were unfairly blamed for a health care system that was becoming untenable: "I fear greatly ... that immigrants could become a scapegoat if we are not careful, and that we could use them as a way of saying, look, we're doing something to keep health care costs down, because we're not letting immigrants come in and they would steal and use all our health care resources. I don't think that is a fair characterization."[171] In fact, it wasn't. As Abu-Laban explains, the creation of a discourse of enemies and scapegoats is one among the many consequences of neo-liberal prescriptions.[172] At the core of these prescriptions, there is the effort of cutting the welfare state and implementing a strategy of cost savings and austerity. In order for such measures to be accepted by the general population, it is imperative to switch the focus from government decisions of surgical reduction in spending to external threats endangering the little that is left. As long as people panic about the arrival of hordes from the outside, they won't notice the real theft going on inside.

In choosing to present opposite perspectives, I am not suggesting a clear-cut division between good and bad guys. As seen throughout the chapter, several political figures voiced their concerns over the years, questioning policies that appeared discriminatory towards immigrants with disabilities. Notwithstanding the existence of these voices, the fact remains that the medical admissibility clause constantly received the support of a majority in both Houses. Concerns for the threat that immigrants with diseases or disabilities could represent for Canada were shared, with small variations, by members of all political parties. On 2 February 1994, speaking of Canadian policy of immigration, Mr Art Hanger (Reform Party, Calgary Northeast, Alberta) noted: "The government in its red book states that we must take humanitarianism and compassion into account in our immigration policy. We are already being more compassionate than any other nation in the world. Is it not fair to demand that this compassion be mated with practicality and a consideration of the other needs in the country?"[173] Echoing these concerns, Mr Sergio Marchi, minister of citizenship and immigration, remarked, "while our system is second to none in the world, its resources are limited and involve the provinces."[174] The goal was trying "to balance compassion and the whole question of being fiscally responsible in terms of a viable health care system across the country."[175] It was unfortunate that such a balance was never achieved. From the early to mid-1990s, with a public mood "decidedly anti-immigrant,"[176] Canada reaffirmed its commitment to attract only immigrants considered economically viable, implementing measures aimed at reducing the number of those accepted under the family class while increasing the number of immigrant workers.[177]

The confrontation between advocates of a more humanitarian legislation and those who saw themselves as the defenders of the Canadian system reached its apex in the early 1990s when HIV and AIDS – a problem that Western countries had ignored for years, relegating it to the Third World and to the rejected of society such as homosexuals and drug users – abruptly came knocking at the door. Suddenly, Canada and other industrialized countries realized that so-called productive members of their societies were not immune to HIV and AIDS. An effective strategy was required to limit its impact. While it would have been wiser to concentrate on prevention and education, these are long-term goals that are unhelpful in diverting immediate fears. Better to find an easy target to blame. Once more, immigrants became the scapegoat for those who wanted to believe the plague of the century could be stopped at the border with a blood test. On 15 April 1994, Mr Art Hanger complained that no HIV test was required for the purpose of immigrating to Canada and asked, "Why is the minister's department not testing each and every immigration applicant for HIV and why are we letting these people into Canada?"[178] He was outraged, since "the minister cannot deny that HIV infected immigrants are a threat to our already overburdened health care system."[179] The dominant narrative around immigration continued to rely on the distinction between nationals who deserved rights and benefits and outsiders who wanted to abusively appropriate what was not theirs.

On 23 September 1994, Mr Hangar brought a motion "to require the regulation of HIV testing for all applicants for immigration and the barring of those who test positive for HIV and AIDS from immigration to Canada."[180] The rationale was that "should someone come to Canada infected with HIV, we the taxpayers are looking to a minimum cost of $200,000. That is a minimum cost to treat each patient until death. That is a cost we cannot bear."[181] He warned his colleagues that "the voting against this measure will cost votes"[182] since, according to the results of a poll conducted by the immigration association the previous summer, the majority of Canadians contacted (no mention was made of the criteria for selection or the language in which the question was framed) favoured the change. If "acting on the will of the Canadian people" was not enough, "there is also the matter of doing what is best, doing what is right and doing what makes sense."[183] The only prospect for eliminating AIDS in Canada was "preventing those who carry this disease from coming into Canada."[184] Accused of discriminating against foreigners with HIV, Mr Hanger responded that his priority was protecting Canadians' interests, "We would be doing a grave injustice to our electorate if we were to hold any priority above its protection and the protection of the services the government administers for it."[185] In a speech reminiscent of the one made almost fifty years earlier by Mackenzie King, he noted:

> Canada has no moral burden to accept everyone ... We as legislators have a duty to ensure that only the best, only the most fit and only the most productive come to the country as immigrants. We have a duty to protect our constituents ... When implemented the motion would be a significant step in the war on AIDS. No one loses. Everyone gains. The people of Canada gain increased safety and a lowering of the burden on our medical system. Legislators gain by voting on an overwhelming popular initiative ... Immigrants gain by having the status of their health thoroughly checked.[186]

The MP concluded: "While we will stop them from coming to Canada as immigrants, at the same time we could very well be providing an invaluable test and invaluable information to hundreds or thousands of people who may not know they are infected. We are not doing a disservice to those who are infected since their fate, I am afraid, is certain. However, we could unintentionally be doing a service to those who are infected but do not know it."[187] With an Orwellian twist, Mr Hanger was able to present his recommendation as a gift to those who were going to be rejected.

The task of answering Mr Hanger's concerns was left to Ms Mary Clancy, parliamentary secretary to the minister of citizenship and immigration. Ms Clancy sketched a succinct history of the country's stance on medical admissibility. She began by recalling the Immigration Act of 1952, whose objective was "protecting public health and safety" and which "listed various diseases and deficiencies that in themselves constituted sufficient ground to deny someone admission to Canada."[188] The act proclaimed in 1978 marked "considerable progress over the old act, especially with respect to medical grounds for exclusion" and introduced two relevant changes: "First, the criterion of excessive demand was added. This measure was intended to protect the universal health insurance system that had been created nine years earlier. It was designed to protect the system from becoming overrun by people who had not paid into the system. Second, the list of illnesses and deficiencies that automatically made a person ineligible was eliminated. Inadmissibility was now decided by medical officers."[189] In her summary, Ms Clancy failed to mention that while a medical officer might be able to assess if someone was going to represent an excessive expense, he/she was hardly in a position to determine other positive contributions that person would make. Furthermore, the concept of excessive demand was left up in the air. In discussing a provision affecting people's lives, it is problematic that such a definition is not included in the act but merely mentioned in the accompanying regulations.

The issue continued to receive attention in the following months with contributions both supporting and opposing Mr Hanger's motion. Among the opponents, Mrs Pauline Picard (Drummond, Quebec) from the Bloc Québécois,[190] noticed: "In his September 23 speech the Hon. Member continues to surprise us by saying: 'When implemented the motion would be a significant step in the war on AIDS.' This shows how little my colleague knows about AIDS. AIDS is an international plague that hits indiscriminately without sparing any society, culture or country."[191] She went on: "We in the Bloc Québécois reject this attitude of denigrating and attacking everything one fears or does not understand; of closing our minds instead of opening them; of telling Canadians: 'Let us keep our heads in the sand and maybe when we stick our heads out again, the AIDS problem will be gone and we will be spared.'"[192] Also opposing the motion was the Liberal member from Thunder Bay–Atikokan, Mr Stan Dromsky, who noted that "current Canadian immigration policy focuses too much on a person's disability and fails to take into account his or her ability to contribute to society."[193] In fact, though the medical opinion is meant to be just part of a general process of assessment which rests with immigration authorities, it often ends up being the only one considered.

Among supporters of the motion, Mr Keith Martin (Reform Party, Equimalt-Juan de Fuca, British Columbia) remarked: "It is expensive to treat somebody who is HIV positive. They do have a series of blood tests ... and we give them medications ... Due to better drugs and better treatment modalities and prophylactic treatments we can use this material to lengthen people's lives. This actually increases the cost to our health system, one that I would say is falling apart at the seams, one that does not have any money."[194] What a true misfortune for taxpayers that medical science had progressed so much! The issue at stake did not merely concern persons with HIV and AIDS, but all those deemed not good enough for Canada. In supporting the motion, Mr Philip Mayfield (Cariboo-Chilcotin, British Columbia) of the Reform Party recalled how Canada had historically dealt with diseased immigrants coming to its shores, forcing them into a quarantine period on the Quebec island of Grosse-Île. Oddly enough, a dark chapter in the country's past came to be celebrated as, in the words of Mr Mayfield, the place "served the purpose of protecting the Canadian population."[195] Mr Mayfield insisted that protection of Canadians "must remain the guiding principle for our immigration officials today."[196] Aside from physical protection against infections and lethal diseases, "There are other factors to consider. Canada's taxpayer funded health care system is available to all citizens who want to use it. This is not the case for most other countries. It is conceivable that individuals knowingly infected with this virus could come to Canada because we have a publicly funded and accessible health care system."[197] The statement assumes that people's decision to move rests exclusively on consideration of economic convenience; according to most findings, however, there are multiple factors leading individuals to search for a new home for themselves and their families in a different part of the world. Migrating to another country requires courage and results in economic, social, and emotional difficulties. Yet the courage immigrants demonstrate gets systematically dismissed within the dominant narrative and replaced with a depiction of migrants as abusers, bogus claimants, and potential usurpers of benefits belonging exclusively to Canadians.

Whereas there was disagreement along party lines on the content of the motion, MPs were on the same page when it came to reject persons who could present an excessive cost to the taxpayer. Even those opposing the motion did so because they were uneasy with barring a category of people and naming a specific medical condition, though they agreed that it was inconsiderate to accept immigrants who had been assessed by a medical officer as being a drain to the economy. Liberal member Mr Sarkis Assadourian (Don Valley North, Ontario) noted, "the government has a law already in place so medical officers can decide who can be and who cannot be admitted to Canada on medical grounds."[198] Mr Michel Daviault (Ahuntsic, Quebec) from the Bloc Québécois quoted in its entirety section 22 of the regulations establishing the criteria to determine medical admissibility:

> For the purpose of determining whether any person is a danger to public health or to public safety or might cause excessive demands on health or social services, the following factors shall be considered by a medical officer in relation to the nature, severity or probable duration of any disease, disorder, disability or other health impairment from which the person is suffering, namely: any reports made by a medical practitioner with respect to the person; the degree to which the disease, disorder, disability or other impairment may be communicated to other persons; whether sudden incapacity or unpredictable or unusual behaviour may create a danger to public safety; whether the supply of health services that the person may require in Canada is limited to such an extent that: the use of such services by the person may reasonably be expected to prevent or delay provisions of those services to Canadian citizens or permanent residents, or the use of such services may not be available or accessible to the person; whether medical care or hospitalization is required; whether potential employability or productivity is affected; and whether prompt and effective medical treatment can be provided.[199]

Most disagreed with explicitly targeting immigrants with HIV (indeed, the motion did not pass). Nevertheless, as stated by Mr Ed Harper

(Reform Party, Simcoe Centre, Ontario), consensus was achieved on the basic fact that "immigration should be a benefit to Canada and not a threat to public health or indeed the economy."[200] Again, the system was not willing to reconsider the predominant narrative of a generous country under assault by foreign abusers.

The subject of immigrants with a disease or disability coming to Canada and using health services that were already strained was of paramount importance for federal politicians. I am not denying the possibility that a few people might try to enter the country because they are in need of services unavailable to them in their country of origin. What I am questioning is why such an occurrence is considered scandalous. This problem came to the forefront of discussion in the mid-1990s when a Polish man with HIV was accepted as refugee and, on national radio, admitted coming to Canada because he was in need of medical services. Members in the House of Commons were outraged. On 20 February 1995, Mr Philip Mayfield condemned the government for allowing the individual to "take advantage of our over-burdened health care system" while "thousands of Canadians are waiting in line to use the system they have been paying into for years."[201] Hon Sergio Marchi explained that the person had been accepted as part of the refugee stream not because he was HIV positive but because, due to his sexual orientation, "there was a well-founded fear of persecution."[202] The minister clarified, "It is not a question of being HIV positive. Each individual case must lay before the board [Immigration and Refugee Board] a well-founded fear in terms of a social group persecution."[203] What was left out was any serious discussion on whether it makes any sense for the Canadian immigration system to assess the reasons behind individual migration decisions. Any assessment is necessarily partial given that human beings hardly act under one impulse but make decisions based on a wide array of different considerations. If Canada is within its rights in accepting immigrants for the economic benefits they will bring, why aren't immigrants in their rights when choosing Canada for the economic advantages they will gain once in the country? Furthermore, rejecting people because of their health condition appears discriminatory and against the principles embodied in the Charter. Last but not least, why does Canada believe that only certain immigrants are valuable because of their potential for economic contribution? As remarked by Paul Hunt, "we should not accept this devaluation of ourselves, yearning only to be able to earn our living and thus prove our worth. We do not have to prove anything."[204]

Politicians have traditionally attempted to justify the medical admissibility clause by highlighting the problems that would result for the citizenry if new users were allowed to benefit from an already small pot of money. A more pertinent discussion would rather focus on two core issues: on the one hand, the narrative portraying immigrants as undeserving abusers of a system created for and to the benefit of nationals,[205] on the other, the way disability continues to be perceived, both within and outside our borders "as a problem for the state, for communities, and for individuals."[206] Canada's dilemma is not whether persons with diseases or disabilities should be allowed into the country, but whether Canadian society is able to satisfactorily address the two aforementioned issues. We need to start looking at immigrants as human beings who, collectively, have done far more for this country than this country has ever done for them. Canada would have never developed without the work and contributions of immigrants, nor would it be able to survive today if not because of the continuing influx of more immigrants. We must question the dominant narrative of a generous and selfless country seeking to help millions of miserable souls across the world and start recognizing that Canada has acted in its own interest. We are not the saviours of the world; we are those taking advantage of it.

Despite recurrent discussions, little changed in terms of legislation until the year 2000 when a new immigration bill was unveiled. Bill C-31, An Act Respecting Immigration to Canada and the Granting of Refugee Protection to Persons Who are Displaced, Persecuted or in Danger, known as the Immigration and Refugee Protection Act (IRPA), received first and second readings and was then referred to the Standing Committee on Citizenship and Immigration on 6 June 2000. The act allowed for the entrance of medically inadmissible persons if they were the spouses or children of Canadians. The provision was contested. Mr Leon E. Benoit (Reform Party, Vegreville, Alberta) expressed his concerns that, "if a Canadian marries a non-Canadian, even if that person would be a tremendous drain on our health care system, there would be a blanket acceptance of that person. I have a real concern about that, when Canadians will be bumped from the waiting lines for receiving health services and non-Canadians will be allowed in to be a drain on our health care system."[207] Ms Elinor Caplan, minister of citizenship and immigration, responded: "The reality of what's happening today, Mr. Benoit, is when a spouse is refused because of medical inadmissibility, they are able to appeal to the immigration appeal division at the IRB [Immigration and Refugee Board]. In almost 100% of cases, they win at the IRB on the basis of humanitarian or compassionate consideration ... this legislation ... makes no change in what is in practice."[208] Contested was also the proposed removal of the admissibility bar for adopted children of Canadian citizens. Ms Caplan emphasized that the proposal came out of consultation with the provincial governments since "most provinces always agree that adopted children who were lucky enough to be adopted by a Canadian family, but who may be medically inadmissible because they have a disability should be welcomed and accepted into Canada."[209] Discussion on the subject had already taken place on 29 March 2000, while the committee was discussing Bill C-16, An Act Respecting Canadian Citizenship, when Ms Caplan had reported: "For the purposes of international adoption, the intention is that a medical will be required for information purposes only. That information will be made available to the parents and the province, so they can know the health status of the child before the adoption is completed."[210] The information raised more than one eyebrow, especially after the minister responded affirmatively to a question by Mr David Price (Progressive Conservative, Compton–Stanstead, Quebec) about whether children adopted by Canadians would become automatically Canadian citizens.[211]

The study of Bill C-31 ended in October 2000 at the end of the thirty-sixth Parliament. It was presented again as Bill C-11[212] at the beginning of 2001 when Parliament resumed and received royal assent on 1 November. The two bills differed in organizational and technical aspects, though there were some substantive changes.[213] Bill C-31 had intended to remove for the first time the bar on admission for individuals who applied as sponsored spouses and dependent children. Several social organizations deemed such a change insufficient and the proposed legislation was criticized for its shortcomings in meetings of the Standing Committee on Citizenship and Immigration. On 1 May 2001, Mr Laurie Beachell, national coordinator for the Council of Canadians with Disabilities, explained that, in line with its mandate, "CCD has a long interest in the Immigration Act and a long-time concern ... that is the discrimination that exists within the act that can prohibit the immigration of individuals with disabilities to Canada."[214] With regard to excessive demand, the organization's stance was that it had been short-changed for years by a government unwilling to address discrimination against disabled persons. Mr Beachell noted: "Our present law would prohibit someone like Stephen Hawking from becoming a Canadian citizen ... Our present law has a stereotypical attitude toward people with disabilities that says they do not make a contribution to society and are a drain upon society ... the Charter of Rights and Freedoms ... prohibits discrimination based on physical or mental disability, yet the Immigration Act continues to discriminate on this basis."[215] Referring to the proposed amendment intended to remove the admissibility bar for spouses and children of citizens or permanent residents as well as refugees, Mr Beachell maintained that while the organization was pleased with the amendment, "it does not go far enough. It should be removed from anyone immigrating to

Canada."[216] He contested the premises upon which the legislation rested: "Our law seems to be based on an attitude that says those with disabilities are not contributors; they are just takers. It is based on an attitude that says people with disabilities will not make a contribution to our society and somehow their demand upon health care is something we cannot bear."[217] The belief that disabled persons were more costly than other immigrants was false since "many of those who immigrate to Canada may put a demand on our health care or social system because of a variety of medical reasons." For instance, "a major business individual may require quadruple bypass surgery within five months or five years."[218] Also debatable were the notion of excessive demand and the qualifications of those appointed to determine it. In fact, "What is an excessive demand? How do you determine this? What education or training is given to people in the field to ensure that the traditional stereotypical attitudes are not what determine who can or cannot come to Canada?"[219] The IRPA Issue Paper 4 referred to the proposed change as having a minor financial impact on the provinces and territories. If that was the case, "Why would we continue to keep the prohibition in place for other people?"[220] No answer was recorded.

On the issue of ministerial permits, the CCD was convinced that "ministerial permits are an abuse of power and are a way of evading having to amend the law to bring it in line with the charter." They perpetuated discrimination among immigrants since "those people who are able to give their cases a high enough profile get a ministerial permit. But that requires community organizations, media, etc."[221] Only people with knowledge of the system or enough money to mobilize public opinion got a chance of receiving a permit. Furthermore "stereotypes exist in our society significantly still. Over half of the complaints to human rights commissions across this country are still on discrimination on the basis of disability. Half the complainants among the existing citizens of Canada still face daily discrimination within our society even though the law prohibits that."[222] What message was the act sending to citizens with disabilities? Mr Beachell had no doubt that many of them felt like "lesser citizens" since "if we weren't born here, we wouldn't be able to come here. So current citizens in Canada are devalued just by the wording of the act as well, because that act continues to provide a devaluation of people with disabilities."[223] While devaluing disabled immigrant applicants because of their disability does indeed send a powerful negative message to disabled Canadians, I believe the discourse is more complicated. In fact, and despite their disabilities, Canadians are never equated with outsiders. The national discourse presents them as inherently belonging to the Canadian family and, accordingly, deserving of the rights and privileges that accrue from that belonging. Whereas Canadians with disabilities are certainly on a lower level in the national hierarchy, they are still part of it. Outsiders have never been.

Several of the points raised by the CCD informed the speeches of representatives from other organizations. On 5 February 2002, the committee heard from Ms Alana Klein, research associate for the Canadian HIV-AIDS Legal Network, an advocacy organization promoting the human rights of people living with HIV-AIDS in Canada and throughout the world. While pleased to learn that under the new act persons with HIV-AIDS were not considered a threat to public health anymore, Ms Klein expressed her concerns with respect to the definition and assessment of excessive demand. While the usual "period over which expected costs will be considered will be five years ... it can be extended for up to ten years in the case of chronic illnesses."[224] That was troubling because "a ten-year projection period is inappropriately long, especially in the case of HIV-AIDS, but also for many other illnesses. The costs for treatment are extremely variable over time. This is a result of medical advances and marketplace considerations, which are constantly in flux, such as the prices of drugs, which are changing all the time. We are concerned that projecting beyond the five-year period would likely be inaccurate."[225] Applying different standards to persons with different diseases was concerning as "having a ten-year projection period for some diseases and a lower projection for other diseases does raise some constitutional issues. Pending charter scrutiny, this could be considered discrimination based on disability."[226] Echoing Mr Beachell, Ms Klein argued that the definition of excessive demand, while overemphasizing "the expected costs a person would be expected to impose on health and social services,"[227] ignored the contributions the person would make to society. She argued: "If the goal is really about protecting the public purse, then the whole public purse has to be looked at – not just what is being spent, but also what's being taken in. In addition, the network submits that non-economic contributions should be considered."[228] It remains problematic how the public discourse developed around disabled immigrants continues to be anchored to concerns about the depletion of money set aside for Canadians, rather than taking into account the effective monetary and non-monetary contributions of those same immigrants.

The next day, in order to obtain clarification on the definition of excessive demand, Joe Fontana, chair of the committee, asked for some background from Mr Brian Gushulak, director general at the medical services branch in the Department of Citizenship and Immigration. Mr Gushulak recalled that the concept was introduced in the 1976 act and regulations and that "historically, the number of individuals who have been refused for excessive demand represents less than 1% of total applicants. The levels vary depending upon the year, but in absolute numerical terms, we are talking about 2,000 to 4,000 people."[229] He admitted that no data was available on the demands immigrants placed on the health system according to their age or gender. The debate flared up when some committee members commented that judging the potential cost of an individual on the health system on the basis of the average cost for all Canadians was nonsensical. Fontana pointed out that a more equitable system of measurement would have been age-specific since "if you used an average basis, I think you are going to penalize older people, in my opinion, or maybe even younger women, whose utilization tends to be a little more than that of younger men."[230] Unfortunately, the suggestion was never implemented.

Tracking down the full debate is complicated by the fact that all meetings of the committee took place in camera after 13 March 2001, and therefore the public is prevented from knowing what was discussed behind closed doors. Still, the results of said discussions are embodied in the IRPA and are reflected in the way immigration is currently handled. It did not help to keep the debate alive that after 11 September 2001 most of the attention around immigration shifted to questions of security and securitization. As parliamentarians repeatedly pointed out, their concerns were limited to what constituents wanted and expected. Why bother with those on the outside? After all, only citizens have rights. There is no duty to look after foreigners. Citizenship remains an exclusive rather than inclusive tool. Derrida's dream of a cosmopolitan city of refuge that would make a "genuine innovation in the history of the right of asylum or the duty to hospitality"[231] is still a dream that no state has ever committed itself to. The reluctance of Parliament to deal with those outside the national boundaries is evident when considering that since the second half of the twentieth century, the number of parliamentary debates on the issue has gradually dropped. Policymaking has become the business of bureaucrats, thus emerging "in the form of regulations rather than legislation."[232] This shift is troubling because, although the Department of Immigration always had considerable influence, at the end of the day "politicians made immigration policy."[233] Even during rare parliamentary interventions, the MPs participating in the discussion tended to be relatively small numbers of concerned people, most of them coming from the ranks of previous ministers of the department or from ridings with a high percentage of ethnic voters.[234] Overall, the issue was far from being a primary concern.

In conclusion, the analysis of parliamentary debates from 1902 to 2002 shows general uneasiness with medically inadmissible immigrants:

parliamentarians avoided the topic whenever possible unless it suited their interests. When discussion occurred, it overwhelmingly rested on the assumption that immigrants were expected to be helpful to the country and that those with a disease, disorder, or disability, whatever these terms meant at different points in time, were not. Either because of unproductivity or the costs placed on Canadian health and social services, they represented a burden and were to be kept out. As many of us do, politicians perceived disability as "a functional problem at the level of the individual."[235] Get rid of the individual to remove the problem. This attitude was shared by members of all political parties, though there were a few voices of dissent. The main point of controversy was on how to proceed in the exclusion rather than on the exclusion itself. Contrasts were often the result of political gamesmanship and blaming the opponent rather than emerging from substantial disagreement. MPs blamed each other to take advantage of the opponent's difficulties, but little difference emerged in the basic assumptions about the non-place of medically inadmissible immigrants within Canadian society.

The analysis has also revealed that over a hundred years, most of the changes to the provision of medical admissibility were in terms of language. Different words were used although no significant shift is detectable in the content of the provision as contained in the various immigration acts passed by Parliament. Aside from economic, social, and political circumstances, devaluation of medically inadmissible immigrants continued to rest on long held assumptions about the worthlessness of persons with disabilities vis-à-vis the able-bodied population. What was unique to and further complicated the situation was that medically inadmissible immigrants were not simply persons with disabilities, but outsiders. This status of foreigner offered politicians the opportunity to pick and choose rather than being forced into passive acceptance, as was the case with disabled Canadians. While the latter, whether valued or not, remained an integral part of the national body, disabled immigrants were either removed from that body or, most often, were simply prevented from entering.

The purpose of this chapter has neither been to present politicians as insensitive human beings, though a few of them certainly were, nor to suggest that their attitudes were different from those of other politicians in Western societies. Canadian legislators were not an anomaly; Foucault observes that biopolitics – "a politics which measures and regulates, constructs and produces human collectivities through death rates and family planning programs, *health regulations and migration controls*"[236] (emphasis added) – is at the origins of capitalism and the nation state.[237] Control of the population is essential to the power of the state and its apparatuses.

Immigration is one such field where this control is easily exerted insofar as immigration is predicated on the principle of selection. Whereas controlling the national population is possible but could potentially inspire an adversarial response on the part of those controlled, no response is expected from those who are not here. In this sense, immigration is a discourse of the state with itself and does not require any negotiations with the object of the state's actions and decisions. My intention in this chapter has been to provide the reader with some insights into the dangers inherent in biopolitical practices, particularly when it comes to the regulation of medically inadmissible immigrants: Whose lives are worthy and whose lives are not? And how are these practices impacting on the notion of citizenship, from entry criteria to membership and participation?[238] This is not merely academic speculation as the issues at stake concern the entire society, a society that prefers looking at persons with disabilities as a burden and at immigrants as outsiders who must prove their worth, a society that assesses potential members exclusively on the basis of economic utility. Politicians were and are outcomes of that society. Let us now move a step further and investigate how the same issue was debated in the press. This time the focus will be on the opinions shared by a number of correspondents, editorialists, and ordinary citizens through various articles, letters, and editorials that appeared in the *Toronto Star* and the *Globe and Mail*.

3 Medical Admissibility: *Toronto Star* and the *Globe and Mail,* 1902–1985

The previous chapter examined the discourse surrounding medical admissibility among federal politicians between 1902 and 2002. In the conclusion of the chapter, I argued that politicians' uneasiness with the subject, far from being an anomaly, was the result of societal understandings of disability and immigration. I now explore the discourse circulating among sectors of the public throughout the same period. The present and following chapters examine articles, letters, and editorials that appeared in the press. This chapter deals with the pre-Charter period, while the following chapter considers discussions taking place after the passage of the Charter of Rights and Freedoms. As discussed in chapter 5, the Charter represents a milestone (and let's be clear that milestone does not mean panacea) in the pathway for recognition of persons with disabilities. Its impact has been felt throughout Canadian society. The analysis of selected newspapers reveals, at least to some degree, how the discourse surrounding immigration of persons with disabilities has changed since the passage of the Charter. Journalism can be a tool to understand the impact that certain issues have on society. Jay Rosen argues, "the news is always getting mixed up with our public and popular cultures, returning 'us' to us with all of our excesses and discontents, but also setting out a pattern, amplifying a tone, and inviting particular behaviors."[1] The press does not simply reflect on certain events, it "is an active agent in public life."[2] My goal is to investigate how the press has contributed to the creation of a public discourse around immigrant applicants with disabilities.

The material analysed in this and the following chapters comes from the *Globe and Mail* and *Toronto Daily Star.* The two newspapers were published throughout the period investigated though, at times, under different names/owners and with differing scales of circulation. The *Globe and Mail* was created in 1936 after the merging of the *Globe* with the *Mail and Empire*; all articles published previous to this date were retrieved from the *Globe*. Both papers started publication on a city scale, the *Globe* in 1844, the *Toronto Daily Star* in 1892. In time, they expanded their circulation to cover the province, though only the *Globe and Mail* has reached a national scale of circulation. The paper has defined itself as Canada's national newspaper since the 1900s; however, this was more of a wish than reality until 1938, when the Trans-Canada Airlines began shipping papers across the country.[3] Despite the *Globe*'s effort to reach out to the country, Ontario (and in particular Toronto) remains the focus in both publications. The selected sources are English-Canadian newspapers and therefore are silent on the outlook of French Canada. The choice is neither meant to suggest that the experiences across provinces were identical to those in Ontario nor to dismiss as irrelevant different understandings across the country. It simply rests on the fact that Toronto has historically been the main gateway city for those seeking entry into Canada.[4] Throughout the 1950s and 1960s, over half of the total number of new arrivals settled in Toronto.[5] On 1 July 1967, the *Toronto Daily Star* reported, "about 55 per cent of all immigrants head for Ontario – and the vast majority of these come to Toronto."[6] The trend is unchanged for the years preceding and following 1967. In 2001, foreign-born immigrants represented 44 per cent of the total population in the Toronto's census metropolitan area.[7] The 2006 census confirmed that in the five years from 2001 to 2006, Toronto received 40.4 per cent of all newcomers to Canada.[8] Besides, central Canada has been at the forefront in the development of the popular press within the country since the 1870s[9] and was therefore considered a good place to start an investigation of the kind conducted in these pages.

In chapters 3 and 4 I analyse all items containing references to immigrants and immigration from 1902 to 2002. They were retrieved by online searches of both newspapers' archives. The selection was then restricted to relevant pieces that did explicitly mention medical admissibility. A total of 358 items was examined, 186 of them from the *Toronto Daily Star* and 172 from the *Globe and Mail*. The research method draws on critical discourse analysis and consists in investigating the messages spread around by the press through the analysis and examination of the structure of communication. It focuses on quantitative as well as qualitative aspects, the former being mostly interested in measuring frequencies (how many articles are written on a topic, how many lines or columns are given to a story, etc.), the latter looking at the analysis of themes, terms, and expressions.[10] The investigation does not merely look at how the media framed certain issues at different times, it also aims to reveal how the discourse that was developed served to legitimize unequal power relations within society, thus reproducing and validating elites' control over less powerful groups, in this case immigrant applicants with disabilities.

This analysis is political in nature as it attempts to understand the "intricate relationships between text, talk, social cognition, power, society and culture"[11] in order to change the status quo.

Communication is shaped by dominant values and attitudes. Behind newspapers there are interest groups at work to emphasize or hide certain aspects, thus shaping the narrative and directing the readership while also responding to readers' expectations and ideological perspectives. As Edward Said notes in his study on the American media coverage of Islam, "'news' does not just happen"[12] but is selected and translated to the audience with the goal of producing and maintaining certain dominant assumptions about reality. Newspapers are products meant for sale; hence, they must provide the material requested by their target market, in the process "promoting some images of reality rather than others."[13] Readers become a commodity, and the media give people what they expect.[14] At the same time, newspapers are also compelled to address other significant issues, even when unpopular with their readership.[15] Such issues are guaranteed at least a limited coverage, thus making newspapers representative of debates and discussions within the broader society. Media, and this includes newspapers, television, radio, and more recently the internet, are both a reflection of and a powerful influence on society, intervening "into our everyday frameworks for making sense of the world."[16] As Ferri and Connor remark, "in everyday places, from living rooms to front porches, bars to barbershops, beauty parlours to bodegas, newspapers are an important part of private and public conversations."[17] On the other hand, whereas the press might be compelled to discuss controversial issues, it does so from a specific perspective, usually the one embraced by the elite. Western newspapers pride themselves on providing realistic and factual coverage, as shown by the variety of opinions embraced, some of them unconventional and critical of the status quo. Yet, regardless of the plethora of perspectives offered, dominant views and interpretations are given priority, in both quantitative and qualitative terms.[18] The press is neither objective nor neutral in its coverage, but it is always attentive to "what is to be portrayed, how it is to be portrayed,"[19] and with what frequency it should be portrayed. No matter how much newspapers like to present their coverage as fair and representative, some voices are "censored, some opinions are not heard, some perspectives ignored."[20] It is up to the reader to navigate the discourse presented by the media and, in the process, uncover what is said and what is purposefully omitted.

The relation between media – in this case, newspapers – and audience is interactive: media constructions reflect as well as influence the values and beliefs systems of their audience.[21] As much as media give space to issues that are of interest to their consumers, they also have the power to channel attention towards certain topics and events. Despite the greater autonomy gained by media in the last century, ties with powerful interests continue to play a considerable role in affecting the message going out.[22] For instance, political figures are allowed constant opportunities to present their perspective to the public, thus setting the framework of intelligibility when it comes to the discourse around policy and policy development.[23] With respect to the changes in immigration regulations and policies, extended discussions concerning immigrant medical admissibility received limited attention in both papers, and the overall coverage was informed by the dominant discourse around the necessity for healthy and hardworking immigrants. In the majority of cases, articles and editorials published on the subject in the *Globe and Mail* and *Toronto Star* were comparatively shorter than those dealing with other topics and tended to be situated in sections of relatively low importance. Articles predominated over editorials and were frequently limited to reporting bare facts, with little space for comments and opinions that could question the dominant narrative. Throughout my investigation, I have found a total of 23 editorials, 9 letters from readers, and 132 articles in the *Globe* and 27 editorials, 14 letters, and 148 articles in the *Toronto Star*. The scarce attention given to medically inadmissible immigrants was expected when considering that the number of these immigrants has typically been small. As a consequence, there was never a vast pool of readers interested in the topic. With a high concentration of different ethnic groups in the country (which resulted from the elimination of racially discriminatory admission policy starting in the 1960s) the attention has mainly focused on changes concerning racial biases and integration into mainstream society rather than medical admissibility.

The reader should also remember that different newspapers have different political orientations and are addressed to a particular readership. With respect to the selected papers, the *Globe* has consistently presented itself as a paper addressed to a mostly well-educated and business audience, while the *Star* has claimed to be a more populist paper aimed at the "little guy."[24] Notwithstanding differences in the approach to political, social, and economic issues, Western newspapers show a clear imbalance in favour of certain dominant views and representations of reality.[25] Furthermore, it should be remembered that at least for the first fifty years considered in this book, newspapers were edited for a public less diverse than the current one. Readers were from a common sector of society, namely those literate people who were willing, because they were interested and had enough time to spare, and financially able to buy a daily paper. Such readership was internally homogeneous and this is reflected in the only slightly different approach taken by the two papers in their evaluation of Canada's interests.

Upholding "A High Standard of Citizenship"[26]

In the early 1900s, limited space was given in the *Star* to discussions surrounding immigration while more references were found in the *Globe*, particularly on the subject of undesirable immigrants. In May 1902, reporting on the debate in the House of Commons occasioned by Mr Sifton's bill to amend the Immigration Act, the paper quoted part of Mr E.F. Clarke's intervention on the need for "definite instructions given the agents in Europe to prevent Canada being made a dumping ground for immigrants unfit for admission."[27] The paper supported the call and, on 22 August, reiterated that "adoption of more stringent regulations is a timely move. Every country is justified in protecting itself against foreigners likely to spread disease or become a public burden."[28] While the preceding decade never saw immigration levels pass the 50,000 threshold, immigration underwent a phase of expansion after the election of the Laurier's Liberal government in 1896. As a result of government efforts to attract immigrants from old sources – Britain, the United States, and western Europe – and new ones in central and eastern Europe, the number of arrivals jumped to 89,102 in 1902, with continuous growth in the following ten years.[29] Despite being beneficial to development, particularly in western Canada, increased immigration brought new economic, social, and cultural problems.

With rising levels of immigration, more space was reserved in the *Globe* to the topic of non-desirable immigrants. As noted in chapter 2, the issue was generating tensions with the United States, and a number of articles focused on complaints from the southern neighbour that unfit immigrants were allowed into Canada and, from there, attempted to cross the border. One of these articles explained that, according to the testimony of Mr Robert Watchorn, in charge of the US Immigration Inspection Service in Canada, undesirable immigrants, many of whom were "diseased" or "idiotic," found an easy way of "coming into the United States through Canada"[30] where inspection and regulations were not strict enough. On 16 July 1903, the editorial "Our Immigration Policy" observed, "That those incapacitated, physically, mentally or morally, for becoming good citizens should be firmly excluded is self evident ... from the standpoint of our own interests an unduly free policy is indefensible."[31] One year later, the *Toronto Star* commented on the disconnect between assurances provided by the Canadian minister of the interior that "the medical inspection of immigrants ... was as perfect as it could be made"[32] and negative reports from US officials. President Roosevelt, in the annual message to Congress, recommended "that no immigrants be allowed to come in from Canada and Mexico."[33] The problem led to a rigid border inspection between Canada and the United States. The editorial "Stop Poisoning the National Blood" remarked: "It is intensely annoying as well as disquieting to read that American immigration officers stationed in Canadian cities are constantly refusing foreigners, who have had no difficulty in getting into Canada, permission to cross into the United States. This means, of course, that they stay in Canada and become a burden and a danger to our population."[34] This situation was unfortunate since "it is all very well to talk about our boundless prairies and the nobility of offering an asylum to the oppressed of earth; but the truth is ... that it is not the 'oppressed of earth' who come seeking an asylum, but the diseased and incapable."[35] The writer considered this occurrence unacceptable.

The goal of keeping out undesirable migrants was shared by the *Globe* in an editorial published on 30 October 1906: "The right of this province or of the Dominion at large to adopt defensive measures against the mentally defectives sent from other countries cannot be questioned. Not only are such persons likely to become a serious public charge which the people of this country are under no obligation to bear, but their presence may be the means of promoting the noticeable and deplorable increase in mental disorders."[36] The paper applauded the government's decision of deporting immigrants with mental defects within two years of their arrival. The measure was considered necessary for the well-being of Canada and in particular Ontario: "The saving to the Province by each deportation is estimated at $2,000, this being based on the actual cost per year per patient, which in Ontario is $135, and the statistical evidence that the average life of a patient resident in an asylum is thirteen years."[37] Almost one year later, the achievements were impressive and the *Globe* congratulated the Dominion authorities for "exercising due care with regard to the quality of these prospective citizens."[38] Reporting that "the immigration policy is working out satisfactorily,"[39] the paper emphasized, "from the standpoint of Canadian citizenship, as well as in consideration of actual burdens on Canadian taxpayers, it is wise and necessary to see that none is admitted who does not conform to reasonable requirements."[40] However, the problem was far from solved and, on 15 November the newspaper gave notice of a meeting of concerned citizens held in Toronto to discuss immigration. At the meeting, resolutions were adopted calling on the Dominion government to "put into effect stringent measures to prevent the influx into Canada ... of mentally and physically diseased ... immigrants" and "adopt a more expeditious method than at present exists of deporting undesirable immigrants."[41]

According to the *Globe*, the government was doing a satisfactory job in preventing indiscriminate immigration of unfit persons. On 19 September 1908, the paper published data showing that "Canada has now a much more efficient system of inspection than the United States."[42] Based on actual record: "In 1906–07–08 there arrived in Ontario 155,460 immigrants. In those years the insane immigrants placed in asylums numbered 130, so that ... the ratio of increase per thousand is only 0.7 ... Moreover, it is important to note that of these 130 admissions sixty-four were deported, leaving only 0.4 per thousand of insane. The United States had an immigration in the year 1906 of 1,085,849, and only deported 307 insane, thus proving that the system of inspection in Canada is much more stringent than across the border."[43]

The exclusion of undesirables was a necessity since "the defectives of other countries are not merely a burden ... but they are apt to perpetuate a criminal or otherwise defective population."[44] For a young country such as Canada, trying to uphold "a high standard of citizenship,"[45] an exclusionary policy was considered a necessity.

Pressured abroad by the United States and at home by several concerned citizens and organized groups, the government was also faced with opposite pressures from the mother country. As legislation concerning admissibility of unfit people became more stringent, Britain saw increasing numbers of emigrants returned to her shores and grew vocal in her opposition. On 15 June 1909, the *Star* reported: "Canada is accused of selfishness in trying to skim the cream of British immigration ... This policy may have its selfish side ... But the attitude of Canada acts as a corrective to a delusion that social problems can be solved by shifting the unfortunates from place to place ... It is a very doubtful kindness to pack off to a colony those who are feeble in mind or body."[46] Among the unfit, paupers were considered on the same grounds with persons with a disease or disability and were subject to a similar screening process. Equating pauperism with disease, the paper quoted the findings of a recent investigation conducted in England: "Pauperism is due to inherent defects, which are hereditarily transmitted. The paupers examined by the committee were characterized by some obvious vice or defect, such as drunkenness, theft, persistent laziness, a tubercular diathesis, mental deficiency, deliberate moral obliquity, or general weakness of character."[47] Rejection was imperative if Canada wanted to avoid the hardship faced by Britain and other European countries: "The information which has been gathered ... emphasizes the necessity of careful immigration restrictions ... these people ... are, in the last analysis, to be viewed as an economic disease, which should be given no hold upon the life of the Dominion ... The countries of Europe are considering the detention of this species. Canada will devoutly hope that if they do not detain them in special institutions, they will at least detain them within their own borders."[48] That hope went unmet and, three years later, Miss Keenan of the Department of Immigration complained that "the country is being made the dumping ground for the scum of European cities, without the slightest doubt."[49] Miss Keenan's comments were made in the context of the decision to deport six undesirable women who were found in Toronto institutions and had come at different times from countries in Europe. According to Miss Keenan, such cases were on the increase as a result of "laxity at the ports of entry."[50]

Troublesome was the issue of the "feeble-minded" whose "moral depravity has eaten and is eating into the vitals of 'Toronto the Good.'"[51] A two-year investigation on the conditions among the feeble-minded in Toronto by Dr C.K. Clarke and Dr Clarence M. Hincks concluded that the situation was alarming. Presenting their findings at the annual meeting of the Academy of Medicine, Dr Hincks attributed the problem to immigration since "54 per cent of some classes of immigrants are feeble-minded."[52] He accused politicians of "permitt[ing] these defectives to enter this country."[53] The Ontario Association for the Care of the Feeble-Minded reached a similar conclusion at its annual meeting held in Toronto in February 1918. A resolution was passed calling on the Dominion government to provide "careful medical inspection in their home locality of all persons preparing to emigrate to Canada."[54] That year, Dr Hincks, in his new role as secretary of the Canadian National Committee for Mental Hygiene,[55] called once again for more accurate medical inspections. As the *Globe* reported on 7 December 1918, Hincks outlined the work done by the committee, observing that "at least 50 per cent of the feeble-minded in Canadian asylums have come from other countries."[56] Immigration, envisioned as the solution to populate western Canada, was becoming a problem.

On 19 December 1916, the *Toronto Star* reported on the discussions held at City Council in conjunction with deliberations on the problem of feeble-mindedness in the city. At the meeting, participants agreed that "we must prevent, if possible, the immigration of the feeble-minded to Toronto."[57] On 29 September 1917, another article gave notice of the recommendations coming out the sixth annual congress of the Canadian Public Health Association. In the closing session, Dr Helen MacMurchy, inspector of the feeble-minded of Ontario, encouraged the federal government to put legislation in place contemplating "the ban on mental defectives entering the country to enjoy the full rights of citizenship."[58] Two years later, Dr A.H. Desloges, general medical superintendent at the hospital for the insane in the province of Quebec, maintained, "there would not be such a large number of foreign-born inmates in our hospitals"[59] if immigration agents had been more careful in the selection process. Accusations that the government was soft with unwanted immigrants were fairly accurate; indeed, up until the late 1920s, Canada's "policy was less restrictive than that of the United States – not because of greater humanitarianism but because money was to be made from new arrivals."[60] According to McLaren and Dowbiggin, a number of powerful interest groups that profited from the arrival of immigrants – railway and steamship companies as well as the Canadian Manufacturers Association – pressured the government to relax immigration restrictions.[61] Apparently, money was dictating government policy.

Immigration of defectives into Canada was gradually grabbing attention, especially at a moment when, after the conclusion of the war, migration from continental Europe resumed. On 27 January 1919, the *Globe* published an article discussing the report submitted by Dr C.K. Clarke, medical director of the Canadian National Committee for Mental Hygiene. In the report, printed in its entirety in the same issue of the paper, Dr Clarke explained, "we have suffered greatly by an influx of defective and diseased immigrants" and "inspection at the ports of debarkation has been a failure."[62] As a consequence, "the country has been forced to assume the maintenance of a large number of people to whom no obligation is due."[63] He requested the federal government to take all necessary steps in addressing the situation, warning, "to build a great nation, we must, as far as possible, use the best materials, not the worst."[64] Endorsing the call for greater supervision, over the years the *Globe* reported interventions of specialists and scholars who encouraged the government to exert stricter vigilance. Among others, Hon Harry Olson, chief justice of the Municipal Court of Chicago, stated that defective European immigrants should be prevented altogether from embarking and suggested "the establishment of psychopathic experts, physicians and inspectors in Europe, who would prevent undesirable aliens from even starting for the Dominion."[65] On 13 November 1920, Prof W.G. Smith, who had conducted an analysis of immigration into the country at the request of the Canadian National Committee for Mental Hygiene, advocated "for the erection of a large immigration station ... along the line of the building at Ellis Island, New York."[66] The message was clear: defectives had to be barred from the country at all costs.

Experts' warnings about the peril of an influx of undesirable migrants, as well as denunciations that little was being done at the governmental level to address the problem, caught the attention of politicians. Measures had to be implemented in response to complaints of inaction. The *Toronto Star* reported: "The recent appointment of a number of extra medical inspectors in connection with immigration into Canada is understood to be consequent on the representations of public health bodies throughout Canada that many imbeciles were being admitted to Canada without proper inspection. The staff of medical inspectors has been more than doubled and every precaution is now being taken to prevent incurable imbeciles from entering the country and becoming a public charge."[67] The paper welcomed the change as long overdue: "Canada has suffered too much in the past from loose inspection to be willing to relax the stringency of its present regulations. There are now too many people in Canada being supported in institutions and constituting a serious economic and social weakness, who have come over as emigrants."[68] The *Globe* observed with satisfaction that "regulations covering the entrance into Canada of European immigrants are more strictly enforced,"[69] though more could be done. On 13 February 1924, the paper quoted Relief Officer A.W. Brown, who asked the town council in Oshawa to take direct action with the immigration department. According to Mr Brown, the department was allowing into the country a "non-progressive type of immigrants ... Some of the individuals and families are physically and mentally unfit for development into progressive citizens."[70] He called for an investigation into people who were "a burden upon the country."[71] Stricter selection was also encouraged by the Social Service Council of Canada, whose delegates at the annual convention held in 1925 accused the federal authorities of conducting an inspection process that was "perfunctory and inadequate."[72] Although "commendatory progress is being made ... by the Federal Government,"[73] the problem of foreign mental defectives was still threatening the fabric of the Canadian nation. In a paper read on the radio during the broadcast of a weekly talk sponsored by the Canadian Social Hygiene Council, Dr C.M. Hincks remarked, "The problem of mental abnormality is made more acute in Canada through indiscriminate immigration. In some Provinces more than half of the insane and feeble-minded have been recruited from countries outside of the Dominion."[74] A few months later, the Canadian Education Association, an organization of Canadian teachers,[75] joined in the call for tightening the immigration barrier preventing "half-wit and subnormal mentalities to slip into the Dominion."[76]

Throughout the first decades of the twentieth century, the *Toronto Star* continued emphasizing the concerns articulated by social organizations with respect to immigration of unfit elements. On 15 June 1921, the paper gave an account of the discussion at the annual convention of the National Council of Women.[77] According to the convener, Mrs Hall, the number of mentally defectives in Toronto and Montreal was steadily increasing and she considered it unacceptable that "in spite of such facts as these, the government spends $1,000,000 each year to encourage immigration."[78] In 1924, in his address to the Toronto League of Women Voters, Dr Eric Clarke urged a more careful watch on immigrants entering the country. Dr Clarke noted how mistakes had been made in the admission of defectives, concluding that "throughout Canada there are communities that are degenerate as the result of improper inspection in days gone by."[79] The focus on mentally defective immigrants was part of a bigger concern with human evolution and the nation's greatness. These anxieties were not exclusive to Canada: all through the 1920s, similar ideas were gaining attention in the United States and Britain. While an organization akin to the British Eugenic Society was only established in Canada in 1930, eugenic theories were widespread among the scientific community and society at large. McLaren notes, "there were few Canadians in the 1920s and 1930s who had not heard some politician or health expert ... employ eugenic arguments and vocabulary."[80] For the eugenic movement, public health, educational reforms, and immigration selection were related components.[81] In his opening address to the meeting of the British Education Society, Prof William McDougall of the Department of Psychology at Harvard University stated that "eugenic is merely social justice."[82] Canada was invited to practice a "selective immigration" that would enhance "the qualities of the stock derived by immigration."[83]

The importance of immigration to the project of nation-building did not only concern scholars and scientists, it was also amply debated among other sectors of society. In this chapter, comments made by representatives of social agencies are not used to capture the views of various organizations, rather, they are examples of the public discourse permeating Canadian society. In 1924 immigration was at the forefront of the discussions animating the Anglican general synod. The issue was further complicated due to the character and mission of the Church of England in Canada. While some members of the synod agreed with Principal Vance of Vancouver who moved a resolution asking the government for careful examination of immigrants, Archdeacon Renison of Hamilton saw the move as a rejection of the church's "primary spiritual mission."[84] Opposing the idea that only the best settlers should be admitted into Canada, he recalled "the saying of the founder of the church: 'I came not to call the righteous, but sinners to repentance.'"[85] Different opinions were also present among trustees of the Toronto Board of Education; at a meeting on 15 October 1925, the board passed a resolution asking immigration authorities for a more careful inspection of new citizens. Trustee Joseph Gordon distanced himself from the motion, declaring that it was offensive to imply "the children of foreign people in this city are less intelligent than other children."[86]

Unfortunately, the few dissenting voices were not enough to challenge the dominant narrative.

Newspapers' reports indicate even less concern about the fate of immigrants among medical professionals. On 27 May 1926, the *Globe* recounted that, addressing delegates of the Ontario Medical Association at its annual dinner, Dr John MacGregor stressed that one of the services doctors were required to perform was educating the public authority on how to prevent the increase of "morons and mental defectives"[87] in Canadian society. Immigration was part of the problem: "We must be very particular regarding the types that we admit. Unfortunately, no small percentage of those finding their way here ... are mental defectives of varying degree, who, shortly after coming, find their way into our hospitals and other public institutions, and become charges on our municipalities. The medical profession can perform a lasting public service by bringing the matter to the attention of the Immigration Department."[88] One year later, the paper published an article written by Frank Chamberlain, secretary of the British Welcome and Welfare League, in which the author complimented the Canadian government for recently establishing a Canadian medical service in Europe with the task of examining immigrants leaving for Canada. Mr Chamberlain wrote: "While this new service will cost the Canadian Government probably $100,000 a year to operate, it will reduce the present hardship, will cut down the number of rejections at the ports of entry, and will reduce the number of deportations within the five-year period."[89] Canadian psychiatrists pressured the government to take decisive action in halting the influx of undesirables. Dowbiggin maintains that such a sustained emphasis on government action, somewhat less pronounced in the United States, was due to a greater belief in Canada's political culture of "the necessity for governments to intervene in health and welfare."[90] These cultural differences between Canada and the United States persist and have affected the development of immigration policies and practices in the two countries ever since.

Deportations in particular were a headache for the government due to high costs and complications in terms of how, when, and under what conditions they were advisable. As seen in the previous chapter, while the process of rejecting immigrants was straightforward, a more persuasive rationale had to be provided in the case of someone who had been accepted into the country and had subsequently become ill or disabled. In 1926, the *Toronto Star* publicized the case of a young Scottish girl, Miss Betty Roy, who was ordered deported after a city relief officer notified the authorities she was a public charge. Mr J.C. Mitchell, immigration officer, justified the deportation order: "She was in the Mountain sanitarium at Hamilton, and these cases cost the city $10,50 a week and the government 75 cents a day. She came to Canada on June 7, 1923, has been in the sanitarium a good part of the time, and in view of the medical evidence we could not do anything else than what we did."[91] Pressures from the public to let Miss Roy stay augmented and a provision was eventually passed establishing that she could remain "if the hospital bill owed to the city in her case were paid and if guarantees were given that she would not become a charge."[92] Once the board of control received an offer to pay off the debt, together with a guarantee from the Daughters of Scotland and other friends of Miss Roy that she would not become a charge, the girl was allowed to remain in the country.[93] In this case, money had made the decision to be generous somewhat easier for Canadians.

Further complicated was the issue of deporting naturalized persons since the law had no provision for their deportation. Nevertheless, the argument was made that Canada had no obligation to take care of these individuals. On 16 February 1928, the *Toronto Star* reported that a change to the immigration law contemplating deportation of insane immigrants after naturalization had been recommended by Dr A.H. Desloges of Quebec during a session of the National Committee on Mental Hygiene. According to Dr Desloges, the measure was fully justified: "Every other country will deport a Canadian, should he become insane ... The fact that he may have become a naturalized citizen of their country makes no difference to them. Back he comes to Canada. But here in Canada when an immigrant has been here five years we assume all responsibility."[94] Neither the *Toronto Star* nor the *Globe* followed up on the debate. The issue was not first-page news, largely because the debate remained on an abstract level, without individual stories catching the readership's attention. Nonetheless, the discussion is worthy of some reflections. Above all, I see it as an antecedent to the neo-liberal rationality dominating current discourse and impinging on the notion of citizenship. The latter should theoretically guarantee inclusion within the Canadian family, with all rights – including welfare rights – associated with it. Yet, inclusion is increasingly assessed in terms of economic participation rather than political standing.[95] In the case at hand, naturalized citizens could be deported, and therefore excluded from the Canadian political community, due to their economic unproductiveness, their access to the rights associated with citizenship wiped away by a predominantly economic argument. The dominance of a neo-liberal discourse is today reaching its apogee with the new notions of "citizen worker" and "consumer-citizen"[96] becoming part and parcel of the mainstream lexicon and recentring the citizenship discourse around economic productivity. While claiming to be both degendered and deracialized,[97] this discourse is nothing of the kind as it takes advantage of the gendered, racialized, and ableist structure of Western societies to position women as well as racial groups and disabled individuals as expendable and precarious workers rather than legitimate citizens with rights. The debate developed in the late 1920s around deportation of naturalized citizens represents the dawn of a narrative that relies on economic productivity as the main criterion for citizenship.

If deportations failed to generate much interest among the public, clamour erupted when the McConachie case exploded like a bomb. I discussed the case in the previous chapter. Given the fact that a baby had been rejected and a family separated, both newspapers were eager to give the story a front-page spot to satisfy readers' curiosity. Opinions varied on the appropriate course of action. On 3 February 1928, the *Toronto Star* concurred with officials of the immigration department that the case indicated "the need of medical inspection of migrants in Great Britain."[98] The writer expressed sympathy for the family but refused to blame the department for merely applying the law. If there was a culprit, it was the mentally defective baby who was ultimately responsible for the family's separation: "Until death comes to the child and thus releases the mother, or until it can be placed in an institution, Mrs. Maconachie must remain with it in her native Scotland while her husband and four other children are in Nova Scotia."[99] Five days later, the paper came back to the case condemning "the unfair and unjust reflection on Canada and Canadian immigration methods,"[100] since the department had acted in accordance with the law of the country. On 11 February, the paper reported that three medical officers in Britain had inspected the little girl after her return home, confirming the findings of the Canadian inspectors. According to British doctors, "there is little or no hope that the Maconachie child will ever be normal."[101] Despite claims that no wrong had been committed, public opinion was divided: "Officials of the department ... have been deluged with letters about the case, some for admittance, some against it. A number of women's organizations are definitely opposed to the admission of a feeble-minded child, as creating a precedent and opening the door for others."[102] Other groups were more sympathetic; among them, "the Prisoners' Welfare Association of Montreal," which "started a fund for the family."[103]

Public pressure to have the child admitted under ministerial permit became so strong that the government felt the need to involve the Department of Justice. Absolving his colleague of any wrongdoing, Mr W. Stewart Edwards, deputy minister of justice, handed down his legal ruling: "Under consideration of the sections to which you have drawn my attention, I am of opinion that said section 4 is not intended to authorize the minister to admit to Canada as immigrants persons who are by the other provisions of the act expressly forbidden entry. It appears to me to be abundantly clear that the minister cannot issue any permit under said section except for a limited period of time."[104] The *Globe* offered a similar legalistic approach by noting, "Government officials were ... acting within the requirements of the law in forbidding entry in this case ... The situation is admittedly difficult for both the Government and the household affected."[105] Two days later, the paper accused those with a different opinion of exploiting the situation to slander the government: "The department [of Immigration] is unable legally to make an exception in this case ... Political capital has been made out of the incident by Conservative politicians and Conservative newspapers, but through examination by medical experts, not only in Canada, but since the mother and child returned to Britain, has shown that 'there is not the least prospect of the mentality of Margaret McConachie ever developing to anything near normal,' and the Minister of Immigration cannot admit the child to Canada as an immigrant without breaking the law."[106] To those saddened by the split of the family, the *Globe* provided its own suggestion: "There is nothing to prevent the mother from joining the father and four children in Canada, save that she has not the heart to leave the infant in Scotland. It is felt here that the wisest course in the circumstances would be for the McConachies to return to Britain."[107] Justice for the Canadian government seemed to be achieved with the report of Professor Kirkpatrick Anderson, "described as the best expert available in Scotland,"[108] who examined the child and found she was "an imbecile with a Mongolian feature."[109] Yet, the concept of justice is a relative one. Foucault got right to the point when, in his lectures at the College the France, he argued that laws "are not means of guaranteeing the reign of justice, but ways of promoting vested interests."[110] Governments can be perfectly legitimate in applying the law, yet this hardly assures that justice will be achieved. Law is a human product reflecting the specific societal context of its creation. What is just for someone might very well be oppressive to someone else.

The year 1928 also witnessed less glamorous stories related to medical admissibility. Some of them reached the press, as in the case of Miss Flora McDowall, a Scottish girl who was rejected because she classified as overweight. On 12 May 1928, an article in the *Toronto Star* reported that the girl was "almost as broad as she was tall."[111] According to a statement from J. Bruce Walker, director of immigration in England: "Miss McDowall was certified three C, owing to marked obesity and stunted height. Her weight was 210 avoirdupois; her breadth almost equal to her height. She was not a domestic and never was. She is now unemployed and because of her obesity is decidedly unemployable."[112] The *Globe* took a different stand by succinctly noting, "the overweight Scottish girl whose application for an immigration permit to Canada was refused must have concluded that the attention of the Immigration Department is taken up with little things."[113] The paper did not consider the story worthy of much attention. More concerning was the issue of mentally defective immigrants flooding the country. In an editorial published two days later, the Ottawa correspondent for the newspaper defined as "sensational"[114] the evidence presented by Dr David A. Clark, assistant deputy minister of health, before the Parliamentary Committee on Immigration: "The number of inmates in Canadian hospitals for the insane who were born outside the country shows, without question, need of a more careful sifting of immigrants that there has been in past years ... Non-Canadian-born in Ontario institutions in the years 1924–26 numbered 3,170. Seventy per cent of the hospital inmates in Alberta were born outside Canada, and information from Saskatchewan was somewhat similar. Quebec reported 600 such public wards. Immigrants like these constitute a heavy liability to the country."[115] The *Toronto Star*'s assessment was different. The paper was supportive of the government's handling of immigration. In an article published on 31 May 1928, ample space was given to the presidential address that Dr Weston Krupp of Woodstock delivered in front of the Ontario Medical Association in which the speaker "complimented the Canadian government on its medical policy with regard to immigration."[116] While differing in their assessment of the government's effectiveness in selecting immigrants, both papers agreed on the basic assumption that Canada had the right to select who to welcome and who to exclude from the national community. That understanding was never questioned and continues to remain the basic framework for discussing immigration.

The year 1930 took off with the reignition of the debate surrounding deportation of disabled immigrants already in Canada, a topic that had received scarce attention just a few years earlier. Following mounting pressure from provincial governments, Ottawa began to strictly apply the regulations, and this resulted in increasing numbers of deported persons. When relatives of these individuals started to voice concerns, the news spread like wildfire and the debate acquired major proportions. The first case to reach the press was the one of Mrs Alice Ainsworth, a Hamilton woman with epilepsy who was deported after eighteen years residence in Canada. On 17 October 1930, the *Toronto Star* noted that the deportation had been "properly carried out"[117] since, according to the official record, the woman was already an epileptic when she arrived in Canada, though the medical examination had failed to reveal the condition. "In 1929 she became a public charge on the province of Ontario,"[118] and the province decided to press for deportation. Asked whether five years' residence protected any immigrant from deportation, the commissioner of immigration responded that protection applied only to those who had entered the country legally: "an epileptic cannot enter legally and therefore cannot acquire domicile."[119] Mrs Ainsworth's husband protested the decision and found support in the mayor of Hamilton, Mr Peebles, who "declared it was a shame to separate a woman from her family after she had lived here for eighteen years."[120]

The impact of Alice Ainsworth's case upon the public was magnified by the fact that in the space of days, two similar stories were reported in Ontario, one regarding Miss Naylor of London, the other Alice Barton of Hamilton, both of whom had epilepsy. Miss Naylor was going to be sent to England where she had no relatives, despite the fact that the family assured the authorities of its willingness to care for her in Canada. The article mentioned, "the girl's mother is seriously ill because of the shock that followed the order against her daughter."[121] In the case of Miss Barton, "inquiries by The Star indicated that Alice Barton could not in any way be termed a 'prohibited immigrant,' as she became an epileptic five years after she came to Canada."[122] Furthermore, her parents complained that the deportation had taken place without them being informed in advance, although immigration authorities denied the accusation. Dr Robert Oliver, physician of the Barton family, wrote a letter to the *Globe* blaming the immigration department for "such unwarranted action."[123] According to Dr Oliver, the parents were denied the opportunity of taking care of their daughter who was shipped out of the country "not like a human being, but like an ordinary piece of merchandise."[124] The action "harks back to the days of slavery, and makes one wonder whether we are living in the twentieth century or not; and in my opinion warrants the fullest investigation of the Immigration Department of the Federal Government and widest dissemination before the public of all the information which can be made available regarding all these cases."[125] Once challenged, the department publicly responded that, as in the case of Mrs Ainsworth, Alice Barton was an epileptic well before entering the country since "she had entered Canada illegally, and could not acquire domicile here. Evidence was given that she had received medical treatment in England for epilepsy and the Ontario authorities insisted upon the law being carried out."[126] The process had been carried out in conformity with the law, and the government was absolved of any wrongdoing.

The explanations provided to the public rested on each level of government blaming the other for poor decisions. The federal government claimed that pressures for deportations originated at the provincial level, while the province justified the move because of financial constraints and lack of structures. Mr Cumberland, superintendent of the hospital for epileptics in Woodstock, told the *Star*: "There are hundreds of such cases in Canadian institutions and if all were deported it would mean the release of accommodation that would mean a good-sized additional provincial institution. The accommodation is needed for other cases."[127] In the same article, the immigration department released a statement clarifying that "the department of immigration regards deportations as the saddest activity it is called upon to perform, and except in the case of out-and-out 'scalawags' it only sends immigrants back to the country of origin as a last resort."[128] The next day, in another article, Mr J.C. Mitchell, Dominion government immigration superintendent for the Toronto district, observed that the deportation orders had been carried out under the request of the province since "hospitals are overcrowded, and ... the public should not have the extra burden of caring for cases that broke the laws of the dominion when they came into the country."[129] Ottawa was unwilling to take the blame. Information gathered from an anonymous immigration official suggested that "the interests of the federal government are against deportation, and the interests of the provincial government for it."[130] According to the official: "The department of immigration dislikes sending back people whom they have invited and assisted to come and, besides, it costs money, takes a lot of trouble and draws down a flood of criticism. The Ontario government, on the other hand, rids itself by deportation of a charge on one of its institutions, undergoes no expense and suffers relatively little criticism."[131] Due to the sympathy the case had generated among the public, neither levels of government intended to play the role of the villain.

In light of the general criticism of deportations within Canada and abroad, the *Toronto Star* became concerned that the country's image could suffer. The paper sided with Ottawa in absolving the federal government from any responsibility while emphasizing the role of the province of Ontario. On 31 October 1930, it reported that after being informed of the repatriation to England of two Ontario women, Hon Robert B. Bennett, new prime minister of Canada, had described the deportations as "monstrous"[132] and said that "when the women had been so long in Canada some other solution should have been found."[133] Indicating that Premier Ferguson (Conservative) of Ontario had pressured for the deportations, the paper referred to a previous speech that Mr Ferguson had given in Hamilton, in which he had complained that "a proportion of immigrants had become a charge on the public, implying that the immigration officials under the Liberal administration had been too lax in admitting immigrants."[134] The writer wondered whether any immigrant could feel safe if people were still considered "an undue burden" after two decades of residence.[135] Ferguson's speech had been "an ill-service to Canada, for it is Canada that has to carry the odium."[136] References were made to comments in the press abroad, especially in England; among other newspapers, "The Manchester Chronicle says editorially that ... Canada is shirking its moral responsibilities."[137] Faced with the potential for a negative impact on Canada's relations with the mother country, the paper chose to blame the province.

Regardless of growing criticism, deportations continued and on 1 November 1930, the *Star* informed its readers of the deportation to Bristol, England, of another epileptic woman, Mrs H.J. Vowels, a mother of four who had been in Canada for sixteen years. In the article, Rev W.A. Cameron of Yorkminster Baptist church remarked: "It is a strange procedure that allows that sort of things. Surely a woman who has been here for that length of time has every right to be looked after by this country."[138] Rev T.T. Shields of Jarvis St Baptist church added that "the deportation is a serious mistake and likely to do injury to Canada's good name."[139] Father L. Minehan, a Catholic, was also critical: "I think it is a very unfair thing to deport people after they have lived seventeen years in this country and made connections here ... I do not think that the public generally will approve of the deportation of these women. It is also unfair for the province to have to bear the expense of these unfortunate people indefinitely; but if we must deport, then I think there should be a three or a five-year limit."[140] Asked for a legal opinion, J.E. Corcoran, Toronto lawyer and former chairman of the board of education, commented, "if a person had been here eighteen years I would think that she had a prescriptive right to remain."[141]

From the municipalities' perspective, the issue was not deportation per se, but the lack of surveillance at ports of entry. Thomas Rooney, city relief officer in Toronto, remarked that "there are 50 per cent of the people whom I report to the immigration authorities who should never have been admitted to Canada ... It costs the city over $1,000,000 a year for its hospital relief work and more than 40% of those who received relief were not Canadian born."[142] Dr J.M. Robb, Ontario minister of health, commenting on recent deportation cases, blamed the press and the federal government while absolving the province of any wrongdoing. Dr Robb accused the press of fomenting the public against the provincial government and "striving to blacken the name of Canada in the eyes of the mother country."[143] Ottawa, on the other hand, was not doing enough to prevent the arrival of undesirable immigrants: "it is our urgent duty to see that our Canadian portals are more jealously guarded in the future than they have been in days gone by."[144] The ineptitude of the immigration department had resulted in serious overcrowding in

Ontario hospitals and high expenses for the province: "When I tell you that out of 1,853 admissions to our mental hospitals in 1929, 593 were foreign-born ... you will readily understand that if it were not for this extraordinarily large percentage of foreign-born mental and epileptic cases, nearly all of our Canadian-born who are now awaiting admission to the institutions of their own native province would to-day be receiving up-to-date specialized treatment for the diseases from which they now suffer unrelieved."[145] Dr Robb insisted that defective immigrants were benefiting from services meant for Canadians. Such an argument continues to be adopted by immigration authorities to this day: as Sunera Thobani explains in *Exalted Subjects*, the Canadian state is built on the understanding that Canadian nationality offers a number of benefits, in terms of rights and entitlements, that must be protected from the predatory activities of undeserving outsiders.[146]

While acknowledging the unfortunate situation the families of the deported were in, the minister steadfastly maintained his support for the decision, especially in light of the scarcity of services available to Canadians: "And who shall say that the undoubted tragedy of those deported cases is comparable to that of the long line of sufferers now forced through no fault of theirs to wait with what patience they may, their turn for admission to our overcrowded institutions? ... Canada is quite within her rights in repatriating those who by reason of disease from which they suffered previously to coming to Canada are becoming a life charge on the province of their adoption."[147] Mayor Peebles of Hamilton disagreed. Convinced that the province was responsible for initiating the deportations, the mayor argued that the Dominion immigration department had merely applied the law and the press had done nothing more than its job in reporting the facts: "Hon. Dr. Robb is merely trying to shift the blame to the press when he accuses the newspapers of creating an unfavorable impression in England ... It is Dr. Robb and his department who are creating the unfavorable impression by their high handed attitude with reference to these deportations."[148] Other public figures concurred with Mayor Peebles and opposed deportation of long-time residents. Hon G.N. Gordon of Peterborough, former minister of immigration, pointed out, "the letter of the law has not been violated, but the spirit of it certainly has been completely discarded."[149] Councillor Eastwood of Bolton considered the deportation "another example of the inhuman action of the Ontario authorities getting rid of unfortunate people who, through no fault of their own, become ill in Canada. They think they can dump these poor creatures at Liverpool like cattle."[150]

Dissatisfaction with deportations was not, however, always motivated by empathy towards immigrants; in particular, there was no agreement as to where to draw the line between those who deserved to stay and those who did not. A letter to the editor of the *Globe* written by a reader from Toronto, Mr William Garside, indicates that some Canadians were against deportation of British subjects, while showing little concern for those coming from outside the Empire. Mr Garside attacked the Canadian authorities for giving "this country the worst possible reputation in Great Britain and Australasia."[151] He went on: "The soliciting and assisting of immigrants from those parts of the Empire to Canada has now gone to the other extreme of deporting hundreds of British residents in this country to the part of the Empire from which they came, in some instances ten years or a score of years after arrival here ... it seems as if any pretext is now put forward to send away many residents of this country – for no worse reason than that they are out of work or have become ill."[152] Mr Garside was only willing to acknowledge the rights of British subjects. He invited officials of the immigration department to refocus their attention "by ridding Canada of a lot of foreigners who have no desire to assimilate with others who are here, and who are of such character and nature that no such thing can be thought of. Send a few hundred thousand of these people with unpronounceable names back to Central and Southern Europe and to Russia."[153] A clear-cut distinction was made between members of the Empire[154] and those who remained outsiders in a land that had never embraced them. Length of residency mattered only for those who belonged. Others dissented with this view and were sincerely outraged by a policy of deportation that seemed oblivious to the disbandment of too many lives, irrespective of nationality. An editorial in the *Globe* observed: "The series of deportation, including the kicking out of men and women who have lived long in the country and who have committed no crime except that of becoming ill ... constitutes a record new to Canada, and one for which no apology or explanation seems to be forthcoming ... A little common sense should be applied before the country becomes the laughing stock of the world."[155] Two weeks later the paper stressed, "the wholesale deportations of the past few months have not only agitated Canada but have also inspired indignant protests in the British Parliament."[156] It called for eliminating "the possibility of similar injustices by placing a reasonable time limit on legal deportations."[157] Another editorial remarked: "There is too often a tendency to regard them [matters of deportations] as mathematical rather than human. But the truth is that tragedy stalks behind scores of these human transshipments ... It is time that Canadians awakened to the fact that our deportation laws are brutal, unjust and unfair."[158]

If deportation was a controversial topic, less concerning was the fate of immigrants who were neither on Canadian soil nor part of the British

Empire. In those cases, the plea to reject them was as strong as ever. It should not be forgotten that in those years, Canada showed no compassion for Jews fleeing Austria and Germany and that "Canadian immigration gates remained virtually closed to these tragic refugees."[159] Particularly intense were calls for keeping out subjects who were neither economically productive nor physically and mentally healthy. Members of the medical profession were at the forefront of the campaign for banning "degenerates;"[160] speaking at the Y's Men's Clubs convention, Dr Peter Sandiford, professor of psychology at the University of Toronto, "urged selective immigration for Canada."[161] He defined the practice of taking in British and other European immigrants who were morally, mentally, and physically weak as originating in a "mistaken idea of philanthropy"[162] and added, "it has been a bad thing racially and nationally, this taking care of the insane and mentally ill, and the burden has become an enormous one."[163] For defectives already in Canada, the best solution was sterilization: "There must be sterilization of the unfit if we are to preserve intelligence of the race ... We can't take the lives of these people by means of a lethal chamber, but we have a perfect right to say they shall not perpetuate their kind. There is nothing painful about sterilization. It is a simple operation, not half as bad as having tonsils removed."[164] In accordance with the general feeling on admission of unfit subjects, the Department of Immigration continued to strictly apply the law. On 29 December 1930, the *Star* reported that a Scottish girl, Miss Mary Doherty, on her way to St Catharines to visit her mother, was refused entrance because "of a slight curvature of the spine."[165] Based on the explanations provided to the paper by the Department of Immigration, Miss Doherty "was physically quite incapable of competing in the Canadian labor market, and neither she nor her relatives were in a financial position to guarantee that she would not become a permanent charge on the Canadian public at some early date."[166] Two other cases were reported in 1932 and 1934. In the first instance, the British wife-to-be of a wealthy farm owner, Mr Fred Seabrook, was refused admission due to her mental conditions. Despite assurances he would cover any future expenses, the immigration department in London told Mr Seabrook that "guarantees against becoming a public charge would not secure admission to Canada of a person who failed to pass the medical examination. No immigrant can enter Canada if he or she has been insane."[167] Two years later the British fiancée of John Steele, farmer and astrologer, was also denied entrance because "she has a slight deafness in one ear."[168]

During the following decade few articles on medical admissibility appeared in either paper, likely because the war brought more pressing matters to light, combined with the fact that the movement of people was halted by the difficult conditions imposed by the Depression and by the war itself. Only with the end of the hostilities and better economic conditions was immigration revived and old problems came back onto the radar screen, this time further complicated by new issues of refugee settlement. After more than twelve years of silence, on 15 February 1946, the *Toronto Star* reported a new case of immigrant rejection. Patricia Stowe, the eleven-year-old daughter of a war worker, coming to Canada with her mother and three brothers from Newfoundland[169] to join her father, was denied entrance on the grounds she was of the "Mongolian type" and therefore "likely to become a public charge."[170] Despite a bond put forth by the father together with three of his co-workers and an uncle, the girl was rejected. In the words of an official at the Department of Immigration, "mental defectives of Mongolian type are barred from Canada by statute ... and cannot be admitted to Canada, even on bond."[171] Asked to comment, Mr Stowe told the *Star* correspondent that he refused to leave his daughter behind and bring the rest of the family to Canada: "What kind of father would I be to do that? She would die without her mother. If she cannot come to Canada, then I will have to give up my job here, and return to Newfoundland, though my prospects of getting work there now are slim."[172] Stowe continued: "I came to Canada when Canada needed my work ... It is not right that my daughter should not be allowed to come to Canada to live with me."[173] His comment was a denunciation of the country's immigration system, a system that claimed to be fair and generous while actually resting on the unequal relationship between those who are nationals with rights and entitlements, and those who are undeserving outsiders chosen exclusively on the basis of their utility to the advancement of national interests. It is an indictment of the underlying rationale behind past and current policies and practices in the field of immigration to Canada.

Canada Is "Not in the Welfare Business"[174]

In the post-war years, the country was also faced with the massive arrival of immigrants coming from European countries that had been devastated and impoverished as a consequence of the Second World War. While maintaining the right to choose who should be accepted, Canadians realized that it was hypocritical to deny assistance to those same persons Canada had fought to rescue just a few years earlier. Moreover, as a founding member of the United Nations (UN), Canada had acquired international responsibilities, among them a commitment to help solve the problem of refugees and displaced persons coming from Europe.[175] It was unrealistic to expect these people could meet Canadian health standards after years of malnourishment and dislocation. On 6 March 1952, the *Globe and Mail* commented on the statements made in the Ontario legislature by the minister of health with respect to the unsatisfactory state of those arriving to Canadian shores. Dr Mackinnon Phillips reported that the provinces had incurred a considerable cost in treating displaced persons. The paper observed that

> these refugees, have gone through years of hunger and hardship, which inevitably have weakened their resistance against disease ... What Dr. Phillips proposes ... is that only supermen and superwomen and superchildren should be admitted to Canada. What he suggests is an examination so grueling that many – perhaps most – Canadians would flunk it. Surely, we cannot be quite so fastidious. Surely, we cannot expect the countries of Western Europe to let us pick and choose among their citizens in this manner, taking one in a hundred, and flinging the rest back as so much riff-raff.[176]

While acknowledging that flexibility was required in dealing with refugees, Canada opposed indiscriminate immigration. In the following years, more articles appeared in the *Toronto Star* dealing with rejections of immigrants with physical or mental disabilities. On 11 May 1954, the paper reported the story of a couple close to celebrating their twenty-fifth anniversary who was prevented from reuniting the family when immigration authorities refused entrance to the youngest child, a seventeen-year-old living in Holland. According to immigration officials, the boy, a skilled bricklayer, was "a little slow mentally"[177] and therefore under the class of people barred from Canada. In 1956, Mr Leslie Horan, a British man wishing to immigrate with his wife and children, wrote a letter complaining about Canadian immigration restrictions. His application had been rejected twice "on account of my youngest son, aged 11 years, being a spastic and unable to walk properly."[178] In another letter, a chemical engineer recounted how his wife was rejected because of an error in the examination conducted by Canadian medical authorities in India. The woman had been examined in New Delhi and "declared unfit because she had trachoma in her eyes."[179] After visiting two eye specialists, she was informed that there had been a mistake.[180] Despite requests for re-examination, the appeal was turned down since "according to Canadian medical rules ... she can only be examined after one year and granted a visa if she is all right."[181] Errors on the part of Canadian authorities were not a sufficient reason for immediate reconsideration.

The regulation barring persons with disabilities continued to stand, though a change occurred in the case of applicant families with a medically inadmissible member. On 23 October 1958, the *Star* reported the story of an Irish family rejected because one of the children was epileptic. An immigration official declared that the policy of the department was "to bar the entire family rather than permit the family to be split up."[182] According to the official, families were allowed to enter Canada and leave a child behind only in exceptional circumstances when the department had been satisfied that the child "will be properly cared for."[183] Although stating that the practice was customary, evidence points to the contrary, as shown in the previously discussed cases of Margaret McConachie and Patricia Stowe. As noted in chapter 1, the decision to reject the entire family when one member was found medically inadmissible was only introduced in the 1952 Immigration Act. Furthermore, what was presented as exceptional was instead quite common, especially if the person barred was no longer a minor. On 28 March 1959, a reader wrote to the paper explaining that his son, living in France, was prevented from entering Canada due to arthritis and because he used a wheelchair. The letter contains the first reference I have found connecting disability and the individual's rights, something that will appear frequently in the years following the passage of the Charter. The reader wrote: "The laws and immigration regulations are obviously designed to protect society, but surely it is not their intent ... to wreak more suffering upon an already helpless and agonized victim of the fortune of war. If they are indeed so inflexible, then surely they are inhuman and contrary to the Bill of Human Rights."[184] The public was gradually starting to formulate a more sophisticated understanding of human rights. The Second World War and the subsequent signing of the UN Charter had undoubtedly played a role in these developments.

In 1959, two issues related to medical admissibility resurfaced in the press. The first concerned immigrants to Canada who faced deportation when sick and in need of hospital care. According to Dr Jack Griffin, general director of the Canadian Mental Health Association, the policy had serious drawbacks as threats of deportation prevented many from seeking appropriate treatment "for themselves or for relatives who could be cured."[185] He urged the government to rescind the section. Obviously, the country was far from benefiting from a policy that prevented immigrants from regaining their health and becoming contributing members of society. The other issue was Canada's dismal record in assisting refugees. In the words of a report prepared by *Star* staff writer Ron Lowman, "we are still niggardly with financial aid to care for the refugees, we still bar our door more closely against them than most other nations, and we have lent fewer trained officials than other nations for overseas refugee work."[186] Moreover: "While Canada virtually ignores the plight of the refugees our selfishness is accentuated by our immigration policy of accepting only the healthiest and cleverest European immigrants. From 1955 to 1958, for instance, we received 43,800. These may have included a few refugees – but only very healthy ones. In that same period, it was the little Scandinavian countries which showed compassion by receiving the sick, the halt and the lame ... Yet their resources and space are smaller than ours."[187] The impact of the report on the Canadian government was all but negligible since two days after the publication Immigration Minister Ellen Fairclough (Progressive Conservative) declared that Canada was willing to admit a larger number of handicapped refugees. However, the number of refugees admitted was dependent on the willingness of the provinces to care for them. While the department was "happy to facilitate" acceptance of handicapped refugees, "somebody has to take care of them when they get here – and that's a provincial matter."[188] Mr Leslie Frost, Conservative premier of Ontario, responded that "she [Mrs Fairclough] didn't know what she was talking about."[189] Also critical was the reaction of the Liberal leader, Mr Lester Pearson, who remarked that the responsibility of accepting refugees rested with the federal government and "it was up to Ottawa to "take the initiative" and work out a plan with the provinces."[190] Having joined the UN, Canada – and not the provinces – had acquired international obligations.

The 1950s witnessed renewed interest around rights for citizens with disabilities. Legislation was passed in Parliament in response to calls for greater recognition of disabled persons' rights. The legislation was the outcome of increased efforts by the disability community in organizing and lobbying all three levels of government. These changes benefited the lives of disabled Canadians more than they did those of immigrants with disabilities. In fact, while passing laws meant to better the lives of Canadians with disabilities, Parliament also passed in 1952 a new Immigration Act[191] that "exhaustively listed the prohibited classes and established the right of examination of immigrants."[192] In spite of political inaction on this front, a new climate was developing around issues of disability and it gradually initiated a shift in Canadians' perceptions of immigrants with disabilities. An article published in the *Toronto Star* on 11 September 1964, reported that at the United Church of Canada's twenty-first general council held the day before, "a more humanitarian immigration policy was urged."[193] The council agreed that "handicapped people should be given an opportunity to enter the country" and "urged adoption of the humanitarian principle that the needs of unfortunate people in this world"[194] could be met in Canada.

A mild critique of the language in the admissibility provision appeared on 22 July 1965, in a piece written by Marvin Schiff for the *Globe and Mail*. Schiff condemned those sections of American and Canadian immigration laws imposing restrictions on the entrance of retarded persons. Although convinced that "a state must reserve the right to refuse admittance to persons who are likely to become burdens to its taxpayers," the author criticized "the language in which sections regarding mental deficients are couched," calling it "emotionally repugnant."[195] Schiff quoted Ruth Doehler, former research worker with the Canadian Association for Retarded Children,[196] as saying, "The words idiot, imbecile and moron have not been found in modern medical literature for many years. They are repulsive as well as outdated. The same can be said for the word insane, and medicine has yet to define the word psychopathic."[197] Professor John R. Seeley, former York University sociologist, argued that "we must recognize the damage that is done by the defining process itself, in which a human being becomes very largely what he is said to be as a consequence of what is said about him."[198] Current scholarship acknowledges that discourse plays a significant role in the construction of different identities.[199] The piece also criticized the fact that, despite the existence of allowances in Canada and the United States for the waiving of those sections barring mentally disabled persons, those allowances were often unknown to the public and were applied inconsistently.[200]

The discussion reached the attention of politicians and a group of Canadians was assembled to meet with US Congressman John Fogarty, known as "a strong spokesman for the retarded,"[201] with the purpose of liberalizing the laws. Among those joining the group, a MP (whose name does not appear in the article) offered to prepare a bill recognizing "the dignity of the individual."[202] Another participant emphasized the need to help families that "cannot immigrate for better jobs when they have a retarded child."[203] As a result of the discussions, Mr Hubert Badanai, parliamentary assistant to the immigration minister, told a York South Liberal meeting that "in light of present medical knowledge," the department was planning "to revise its policy on the admission of retarded and mentally ill persons."[204] Meanwhile, sectors of society continued pressuring the government to amend the legislation. On 12 January 1966, at the closing session of the Third Institute on Mental Health Services, the Canadian Psychiatric Association[205] renewed the call for a different method of assessing immigrants. Dr J.D. Griffin, general director of the Canadian Mental Health Association, suggested that "cases should be decided on individual merit and circumstances. The objection to the present act is that it sets rigid rules and categories. Many persons who suffer from cyclical depression, for instance, are extremely creative and have much to contribute to the country."[206] The association agreed that "landed immigrants who become ill after they are in Canada should not be deported because of mental disorder."[207]

In February 1967, the Canadian Mental Health Association brought its request to the Senate-Commons Committee on Immigration. In that venue, Dr Griffin explained that many newcomers who "enter hospital for a short period of time for treatment of a severe emotional depression ... may recover in a few weeks."[208] Often they were suffering from culture shock defined as "the shock of trying to adjust to new conditions, new values, new ways of doing things. Sometimes the shock is too much for them. Sometimes they break down as a result."[209] Maintaining that mental illness was not necessarily linked to unproductiveness, Dr G. Allen Roeher, executive director of the Canadian Association for Retarded Children, presented data indicating that "600,000 Canadians are retarded to some extent. They and their relatives total 2,500,000. However, 75 per cent of the 600,000 are only mildly retarded and can take jobs in industry."[210] The briefs prepared by the Canadian Mental Health Association and the Canadian Association for Retarded Children received scarce attention in the press. Whereas the *Globe and Mail* barely reported what was said in front of the Senate-Commons Committee on Immigration, the *Toronto Star* included a few comments. The paper took a cautious approach, standing by the principle that "Canada as a general rule should not accept immigrants who are acutely ill or who are likely to suffer relapses" and that "the host country is undoubtedly entitled to set minimum mental and physical standards for immigrants."[211] This stance was in line with the white paper on immigration policy that, the year before, had reiterated the need for keeping out people deemed as misfits.[212]

Regardless of repeated promises, little was accomplished in terms of legislation, though some flexibility was registered in the application of the law. In 1968, the *Star* reported the story of a seven-year-old girl from Guyana, Annabelle Lopes, who had arrived in Canada two years earlier and was ordered deported because she "is deaf and has a heart murmer [*sic*]."[213] Luckily, "placing compassion ahead of department regulations,"[214] Immigration Minister Allan MacEachen intervened, thus allowing Annabelle and her mother to remain in Canada. The paper remarked: "We have had frequent examples in past years of how immigration regulations, rigidly applied, have brought disappointment and tragedy to some who had hoped to build new lives in Canada. It is heartwarming to see that the new immigration minister doesn't hesitate to bend the rules for the sake of a little girl."[215] Four years later, the case of Lina Di Carlo, an eighteen-year-old girl who had arrived in Canada in 1966 and had lived since then under a temporary permit because she had polio as an infant and used a wheelchair, received even wider publicity. Considered an undesirable immigrant, Lina was prevented from getting landed immigrant status and therefore unable to apply for a disability pension. Without the pension, she could not pay for "high school and university and eventually obtain a job to support herself."[216] Miss Di Carlo's plea for being allowed to finish school and "support myself with a job"[217] highlighted one of the many contradictions embedded in Canadian immigration policy: it was the unequal treatment received rather than disability that forbade many from contributing to society. The social model of disability has demonstrated that disability is not so much a pathology of the body but a socially produced system of barriers and exclusion that prevents some from successfully integrating and contributing to society.[218] Although Miss Di Carlo's story had a happy ending when Immigration Minister Bryce Mackasey granted her immigrant status, the department reiterated that rules had been bent but those rules were not inherently wrong. As reported in both the *Star* and the *Globe*, a spokesman for the minister explained: "Miss De [*sic*] Carlo has been in the country for five years and in that time she did not become a public charge. She has done well in school and the minister felt, for humanitarian and compassionate reasons, that she deserved landed immigrant status."[219] In the 1970s, while a number of identity groups were fighting for their rights all around the world, Canadian immigration policy confirmed that immigrants could occasionally receive charity, not rights.

In the early 1970s, another story received widespread attention in the press while offering the minister a new opportunity for compassion. On 6 November 1972, the *Star* reported that the Immigration Appeal Board had ordered the deportation of Lynn Hackett, a twenty-three-year-old visually impaired woman from California with a mild form of epilepsy. Ignoring that Ms Hackett had a job and friends willing to guarantee she would not become a public charge, the board declared Ms Hackett inadmissible and ordered her "deported to her native California."[220] In an editorial published in the same issue, Alexander Ross wrote: "In a way, I can sympathize with the three members of the board who made this decision. On a baldly legal basis, Lynn's case was not a strong one and the members must have been mindful of last week's election results, which reflected the public's concern at this country's overly indulgent immigration procedures. But from a humane point of view, the decision to deport Lynn Hackett is disgustingly callous."[221] Ross's comments are interesting considering that just two days later, both the *Star* and the *Globe* published first-page articles announcing that Immigration Minister Mackasey had his office flooded with so many appeals – "letters, telegrams and petitions"[222] – that he "decided to override a deportation order"[223] against Ms Hackett. According to a spokesman for the minister, "the wave of public sympathy ... was a major factor in the minister's decision."[224] While it might appear odd to reconcile such an outburst of solidarity towards Ms Hackett with Ross's statement that the public disapproved of the government's largesse in accepting immigrants with disabilities, I argue that the two behaviours are in fact complementary. The stories recounted in this chapter suggest that the public often assess immigrants in terms of convenience: as long as they do not impinge on our interests, they are not a problem. Things change when immigrants require health care or welfare, benefits we have been taught to regard as exclusively ours.[225] Yet when a sensational case reaches national attention, there is a tendency to show how generous society can be. Compassion in single cases confirms how humane we are without questioning the general rule. It is an exception, but one that makes society look good.

The following year, another story generated requests for the minister's intervention. On 20 June 1973, the *Globe and Mail* reported that Dr Jun Yee Ho, a thirty-two-year-old Taiwanese woman who had been admitted to Canada on a student visa, was to be deported because she "fell within the section covering insane persons."[226] Public calls were soon directed to Manpower and Immigration Minister Robert Andras to rescind the deportation order. The Canadian Mental Health Association, represented by its general director, Dr George Rohn, defined the action as "one more example of the still-pervasive discrimination and stigma and unfair treatment of people who are sick."[227] Public pressure appeared to carry enough weight. The minister refused to intervene under the excuse that "the Immigration Appeal Board has the final say"[228]; Dr Ho was ultimately allowed to remain in Canada after the board agreed to another hearing and "reclassified her under the Immigration Act."[229] A similar happy ending was reached for Miss Susan Baxter, an Alaskan woman who was hired by Alberta's Department of Health and Social Development in Edmonton, but who was refused entry because accompanied by her "mentally retarded"[230] eleven-year-old son. Under pressure from the National Institute for Mental Retardation,[231] Immigration Minister Andras granted the boy a ministerial permit to enter the country calling "that particular section of the law archaic."[232] Dr Allan Roeher of the National Institute for Mental Retardation, remarked, "if a family is good enough to come into this country and make a contribution, then this country is obligated to accept any handicapped member of that family."[233] Mr Paul McLaughlin, associate of the institute, added: "If we allow a family to enter Canada, we should also admit the retarded child who is a member of that family. The contribution that family is going to make is going to far outweigh any expense there may be of having that child here."[234] Dr Roeher's and Mr McLaughlin's comments are significant in the context of the 1970s, when neo-liberal ideas started acquiring dominance in the formulation of social and economic policies.[235] Within the neo-liberal framework, citizenship is tied to market participation;[236] in the case at hand, the economic contribution of the family outweighed the burden represented by the disabled individual and therefore made the exception not only tolerable but also necessary to the country's financial well-being.

The 1970s and 1980s registered several cases settled by granting ministerial permits for humanitarian and compassionate reasons, without the law ever being revised. In March 1975, parliamentarians called on the minister of immigration to allow Hari Charan Singh, a Fijian citizen facing deportation, to remain in Canada. Although "he had no previous history of mental illness and is reported by his doctors to be making a good recovery,"[237] the Immigration Appeal Board ordered Mr Singh deported on the basis of the section of the Immigration Act dealing with insanity. Pressured by MPs, Immigration Minister Andras agreed to "looking at the case to see what discretion I may, or may not have."[238] After consideration, Ron Button, assistant to the minister, confirmed that Mr Singh was going to receive a ministerial permit due to the "extreme humanitarian aspects of the case."[239] Undoubtedly, the department was pressed to relax its rules, especially in cases when the health conditions motivating rejection or deportation were no longer perceived as threatening to Canadians. An editorial published in the *Star* observed: "Immigrants to Canada have to meet health regulations so that Canadians are protected against communicable diseases, and the prospect of supporting people who will be a drain on society. But some of our health rules for immigrants go beyond that ... Epilepsy, mental illness and retardation are problems that separate people from society in an unfortunate way. Our immigration laws confirm that separation."[240] Impatience was growing at a government that admitted the law was archaic and needed change yet seemed unwilling to act.[241] Strong criticism was directed towards the minister when Patricia Meyers and her thirteen-year-old "retarded" daughter were deported from Canada in December 1975. The Canadian Association for the Mentally Retarded defined the decision "a grave miscarriage of justice."[242] The *Toronto Star* commented: "The majority of the retarded can live in the community and ... can be trained and can even become self-sufficient. So why have Patricia Meyers and her 13-year-old daughter, Dina, been forced to leave the country? Yet everyone, from Immigration Minister Robert Andras to a special parliamentary committee ... has said that a new law mustn't prohibit the entry of retarded children. The change will mean some extra cost to society for the special care and training these children may need, but humanitarian considerations outweigh this objection."[243] A *Globe and Mail* editorial blamed the minister: "Would intervention in this case have brought yells of outrage down upon the head of the minister, Robert Andras? We think not. He could well have tempered the law with mercy, at least in the extent of issuing a temporary minister's permit."[244] While far from questioning the assumptions behind the law, both papers were now pressuring for relaxing some of its regulations.

Also outdated appeared the provision of admission against epileptics. The public became aware of the issue when Vivienne Anderson, a Briton married to a Canadian citizen, "turned down an offer of the landed immigrant status she has fought for three years to obtain."[245] After living in Canada under a ministerial permit since 1974, the woman, who had been battling the immigration department for repeal of the provision barring epileptics, was offered landed immigrant status under the pretext that she suffered "not from epilepsy but epileptiform seizures (seizures akin to epilepsy)."[246] Mrs Anderson declined the offer because, as she said in a telephone interview, "I believe the offer was made to try and shut me up."[247] She explained, "If I accept landed status on the terms offered, it will be a one-off decision, which will do no one any good but myself."[248] Mrs Anderson's decision generated a concert of support, further pressuring the government to review the legislation. An editorial in the *Globe and Mail* commented: "The time has come to ask when we will have a Minister of Immigration with the bare minimum of courage necessary to do what he recognizes should be done. The courage needed, after all, is not great, by no means as great as the courage required by Vivienne Anderson, an epileptic who turned down the immigrant status she had been seeking rather than connive at a questionable Immigration Department dodge that would have left the offending law intact."[249] A spokesman for the department dismissed as "sheer nonsense"[250] Mrs Anderson's claim that the offer of landed status was made to silence her; however, Andras conceded that the section of the Immigration Act barring epileptics was "an extreme embarrassment to the Government."[251] Unfortunately, no legislation was passed to address the problem. On 2 February 1976, the *Toronto Star* accused the minister of good intentions but little action: "Immigration Minister Robert Andras is embarrassed by an outdated section of the Immigration Act barring epileptics from emigrating to Canada. But he's taking a too leisurely attitude to changing it. Epileptic immigrants may now enter the country by ministerial permit which, Andras says, gives them 'the same privileges as any Canadian citizen' ... But an immigrant without landed immigrant status is only a guest in Canada, however privileged."[252] In a letter to the editor of the *Globe and Mail*, the Canadian Association of University Teachers called for prompt action. According to its executive secretary, Mr Donald C. Savage: "Changes to the immigration regulations can be made, in other circumstances, with little delay. Would it not be possible, in advance of any major overhaul of the Immigration Act, to introduce a small amendment to remove epileptics from the list of prohibited persons. Such a small, decent act would be widely applauded."[253] The department refused to bend, reiterating, "no special reform dealing with epileptics will be introduced before a general revision of the Immigration Act, probably this year."[254] Meanwhile, Mrs Anderson turned down a second offer of landed immigrant status. In her letter to Andras, she wrote: "I must once again refuse to put my name to the form whereby I would condone the untruths by which you seek to give me landed status because I have obtained public support and have threatened to take a test case before the court, whilst refusing it landed status to others less articulate whose medical condition is not as severe as my own."[255] Mrs Anderson's resolve illuminated the common tactic adopted by immigration authorities of making exceptions in order to keep the system unchanged. Playing on people's desperation to get in, the department accepts those with enough financial, educational, and social capital to publicize their case. Few have Vivienne Anderson's strength to reject such an offer. More important, few have the options Vivienne Anderson had: for many racialized women from the Global South, coming to Canada is a necessity rather than a preference. Throughout my investigation, I have found only two individuals, Vivienne Anderson and Angela Chesters, who turned down an offer attempting to substitute charity for rights. Both women were white with a stable financial position and the guarantee of a decent life elsewhere. Without taking away the courage shown by both women in adopting a principled position, we must recognize that others have never had that same choice.

Although new immigration legislation had been promised and was long overdue, there were related problems to consider. As seen in the previous chapter, MPs were concerned that a new set of provisions alone was not going to change much if the bulk of the decision-making process was left in the hands of bureaucrats who operated through regulations attached to the law and outside Parliament's control. Analogous concerns were expressed in the press. On 30 November 1976, an editorial in the *Globe and Mail* congratulated the government on its intention to introduce new legislation but remained sceptical about the results. A new law was no panacea, especially when considering that "from the principles enunciated in the law to the decisions made by an officer at a point of entry the linkage may be long and sometimes tortuous."[256] The writer continued: "Whether it works well or badly, will depend more on the regulations and instructions by which it is applied than on the bare bones of the principle, sound in itself, that is stated in the law. And if there is reason for misgiving about this bill it lies in the extent to which it leaves all the real decisions to be made, not by Parliament when it passes, rejects or amends the legislation, but by the Cabinet, or the minister – in either case on the advice of the Immigration Department itself – by regulations."[257] The new immigration policy was going to be further complicated by a set of regulations that very few applicants knew. This in addition to the fact that the legislation was not so welcoming to begin with. Barbara Yaffe, reporting for the *Globe and Mail* on the case of a twelve-year-old Hong Kong boy with Down's syndrome, noted that the long-awaited bill discussed in Parliament, "while removing the categorical ban on entry of epileptics, idiots, imbeciles and morons, does not guarantee that retarded children, such as the 12-year-old Hui boy, will be accepted. Exclusion of such applicants would be based instead on danger to public health or excessive demands on health or social services."[258] The provision of exclusion also applied to persons with physical disabilities. All things considered, it was not by accident that the Immigration Act of 1976[259] eliminated discrimination on the basis of race, national or ethnic origin, colour, religion, or sex, but remained silent on discrimination on grounds of disability. As Sunera Thobani explains, the passage of the act coincided with a period of need for immigrant labour.[260] Faced with a labour shortage, Canada realized that economic necessity required putting aside racial considerations. However, due to the widespread misunderstanding that disability prevents people from being employable and productive, there was no pressure to eliminate discrimination against immigrants with disabilities from the Act.

On 29 November 1978, the *Toronto Star* reported the story of Herman George Aviles, who was paralyzed in a motorcycle accident in Ecuador and came to Canada on a temporary permit for treatment in Toronto. Once the permit expired, he was ordered to leave the country or face deportation. After appeals to Prime Minister Pierre Trudeau and Immigration Minister Bud Cullen, the second of the two publicly refused to intervene since "Canadian taxpayers probably would be burdened with Aviles' medical bills."[261] In a last attempt, Mr Aviles applied on compassionate grounds for a ministerial permit, claiming "Canada has better medical facilities and job opportunities than his native country."[262] Mr Cullen declined to act because he did not want to create a "dangerous precedent"[263] since, "anyone needing a particular kind of medical treatment available in Canada but not in his or her homeland would be in a position to come to Canada and seek to remain."[264] Canadians supported sporadic acts of generosity, but they were not interested in the business of rescuing the world's destitute.

The late 1970s witnessed a major case hinting at how immigration was becoming a political tool to gain or lose the support of sectors of the Canadian population. On 8 December 1979, the *Toronto Star* published the story of Neelam Kohli, a visually impaired woman from India who had been in Canada since 1977 under a visa permit after arriving with her mother, brothers, and sisters, all accepted as landed immigrants. Despite guarantees from the family that the girl was not going to become an economic burden, the immigration department ruled Miss Kohli medically inadmissible and ordered her deportation.[265] Immigration Minister Ron Atkey explained that "there are adequate facilities in India for Neelam Kohli's care."[266] Several readers were sympathetic to the girl. A letter by Ron Santana of Ottawa to the *Globe and Mail* noted: "More than 16,000 persons who were already in the country were permitted to become landed immigrants in 1978, on humanitarian grounds. If the department can make 16,000 exceptions to the rule, one more deserving case, such as that of Neelam Kohli, is not going to make any difference."[267] In the *Star*, some readers of Indian background perceived the rejection as a racist attack against the community. In a letter published on 2 January 1980, Veerendra Adhiya, producer of a Hindi radio program in Mississauga, claimed to speak "on behalf of the majority of East Indians – approximately 100,000 in the Metro area"[268] in asking the minister to intervene. Mr Adhiya added, "This is not a threat, but Atkey should know that there are more than 2,000 East Indian voters in his St. Paul's riding and ... deportation of a blind girl from India are surely going to cost him votes."[269] While letters published in newspapers are hand-picked by the editorial staff and therefore tend to reveal more about the newspaper than the general public, they might unveil societal trends. Ferri and Connor note, "Whether the letters that get published are representative of the range of opinion on a particular issue is anyone's guess."[270] Nevertheless, I agree with van Dijk that in a system where ordinary persons are usually prevented from directly participating or controlling discourse, letters to the editor constitute "modest forms of counter-power"[271] that we cannot afford to dismiss as unrepresentative. Mr Adhiya's letter seems to mark a defining moment in the development of immigrants' consciousness. It should be read in conjunction with the fact that immigrant minority groups in the late 1970s–early 1980s were acquiring awareness of their electoral power, which in turn resulted from an increase in the number of visible minorities accepted after the late 1960s following the elimination of racial barriers in the immigration process. Whereas Employment and Immigration Canada reported Britain and the United States as the main sources of immigration in 1968, twenty years later they had been replaced by Hong Kong and India.[272]

Racialized groups began to realize that their growing presence in Canadian national and provincial arenas allowed them some leverage in their dealings with the government. With an election coming in the month ahead, the *Star*'s announcement that "Immigration Minister Ron Atkey promised yesterday to review a deportation order against a blind East Indian woman"[273] makes more sense. The threat of losing the immigrant vote might have proven more effective than a few opposing voices, such as one reader from Etobicoke who wrote: "That blind girl from India ... should consider herself fortunate to have been allowed to remain in Canada for more than two years when she was admitted only on a visitor's permit. Illegal immigrants should be allowed no appeal. Wake up Canadians, and speak up against these illegal immigrants."[274]

Unfortunately for Atkey, a new story came under the public screen a few days later. On 25 January 1980, the *Star* reported on the case of Harjit Kaur, an Indian woman who found out, after joining her husband in Canada, that he had decided to leave her. The discovery had thrown the woman into depression that "swiftly grew into acute mental illness."[275] Mrs Kaur was ordered deported, but the order was repeatedly postponed because she was "too ill to fly back."[276] Nevertheless, Atkey remarked, "Canada was not in the welfare business."[277] The persistence of the immigration department in wanting Mrs Kaur deported to India raised concerns. A *Toronto Star* editorial commented: "The first task of the next minister of immigration ... should be ... to teach the department's officials the meaning of compassion ... no one is saying that Harjit Kaur should be allowed to stay on in Canada forever, but one can't help questioning the unfeeling way the immigration department goes about its work."[278] The attack did not go unnoticed and the minister replied: "Neither I, nor the immigration officers whose duty is to administer the Immigration Act, fail to recognize that compassion and generosity of spirit must accompany the performance of that duty ... Unfortunately there is a conflict between our obligation to uphold Canadian law and the attempt of various persons to enter the country illegally. It is disappointing and disheartening to all of us when a supposedly reputable paper openly supports flouting the laws of our country."[279] The situation became complicated when Harjit Kaur married Ravinder Kumar, a milk store operator. To everyone's surprise, the minister decided "to go along with the 'expulsion order' despite the marriage."[280] Seeing no sparks of light, the newlywed couple appealed, without success, to US President Jimmy Carter maintaining that "in India she [Mrs Kumar] would be in the same position as the U.S. hostages in Iran. Her life and welfare would be in jeopardy."[281] In what resembled a real-life soap opera, the debate was fuelled by the fact that the lawyer representing Mrs Kumar was Mr James Lockyer, "the New Democratic Party candidate running against Immigration Minister Ronald Atkey in the Toronto riding of St. Paul's."[282] Amidst reciprocal accusations and constant media attention, Mrs Kumar was eventually deported on 12 February, after the federal court rejected an application to keep her in Canada due to "poor health."[283] In explaining the ruling, Judge Hugh Gibson said that the court "had no power to restrain the minister from carrying through with the order unless he was committing an illegal act, which he was not."[284] The only comment made by immigration official Ron Bull was a reassurance to the public that "he had heard that Air Canada had prepared a special meal for her on board the Toronto-London flight."[285]

The deportation did little to stifle the debate. John Picton wrote for the *Toronto Star* that Mrs Kumar's departure left behind "a bill for air fares totalling $6,125 – including round-trip tickets for a female escort and a doctor – and other costs to Canadian taxpayers that officials estimate may go as high as $80,000."[286] The treatment committed to Mrs Kumar was harshly criticized. Several readers empathized with Mrs Kumar, though there were also voices of condemnation. Mrs Davidson of Toronto sent a letter to the *Star* expressing her disgust "with all the sympathy given illegal immigrant Harjit Kumar ... She cost the Canadian taxpayers $80,000 for lawyers, hospital expenses etc. She should not be allowed to return to Canada."[287] Once again, the narrative of immigrants scamming the system was revived to make up for the inefficiencies of Canada's social and medical services.

Politicians and the public were not alone in their concern with the kind of immigrants entering the country; in April 1980, after an enquiry into the suicide of a Korean man who had jumped through the window of his apartment, an inquest jury recommended that "anyone with a history of schizophrenia should be banned from emigrating to Canada."[288] Despite having declared to immigration authorities that neither he nor members of his family had ever suffered from mental illness, it was later discovered that the man had previously "spent three months under psychiatric treatment in Korea and two of his four brothers and sisters also had had mental problems."[289] The Toronto coroner's jury asked for all immigrant records to be kept on file in Canada until the immigrant became a citizen and for the screening of parents and siblings of schizophrenics.[290] Members of the health service community criticized the jury's recommendations. Dr Brian Hoffman, staff member at the Clarke Institute of Psychiatry, noted that "there is often a wide difference among countries' definitions of a schizophrenic."[291] He expressed concerns for people coming from the United States where schizophrenia was often equated with "deviant lifestyles" and for those arriving from the Soviet

Union where political dissidents were jailed "under the label of schizophrenic."[292] John Gouyea, spokesman for the Canadian Mental Health Association, stressed that children of schizophrenics had only a one in ten chance of being affected by the disease. Frank E. Cashman and Joel Jeffries of the Clarke Institute of Psychiatry added, "a diagnosis is not always completely reliable" and "some people are mis-diagnosed."[293] Furthermore, the rejection of schizophrenics was not necessarily a benefit to the country: "One need only mention individuals like Nijinsky, who was schizophrenic; James Joyce, one of whose children was schizophrenic; and Vincent Van Gogh, who was possibly schizophrenic, to realize the dangers inherent in blindly following the dictates of the 'genetic' of schizophrenia."[294] No letters from the readership were published on this subject in either paper. From previous cases, it could be inferred that readers were more prone to show support for accepting medically inadmissible immigrants when individual cases rather than general policies were brought to their attention. Being charitable with a single person is easier than questioning the entire system upon which society rests. Proof of this emerged when public support erupted with the *Globe*'s reporting of the story of a Latvian Jew who was denied access to Canada because he was in need of "an expensive kidney operation which relatives in Canada cannot afford."[295] Fortunately for the young man, the article came to the attention of a wealthy benefactor "who agreed to pay $30,000 in medical costs which would cover the kidney transplant."[296] As soon as the government received the financial guarantee, "a minister's permit to enter the country was granted to Mr. Markh."[297]

As the above case suggests, the main issue in allowing persons with disabilities or non-communicable diseases to enter the country was the cost to taxpayers. The introduction of the Medical Services Act in 1968[298] created a national public program securing public coverage of visits to hospitals and doctors together with access to diagnostic services.[299] Whereas initially moderate concerns were raised about the cost involved, wider sensitivity to the money spent developed in the early 1980s when the country went through an economic recession. Only once a guarantee of external financial support was secured, either by independent individuals/organizations or by provinces, the federal government allowed entrance into Canada. This point is proven by the positive reception among federal politicians of the proposal advanced in 1981 by the Manitoba government to start accepting "disabled and handicapped refugees from Vietnam and other Southeast Asian countries."[300] The plan was meant to "allow a refugee to join family members already in Canada despite the fact that he does not qualify for entry because of a physical handicap."[301] Ronald Collet, spokesman for the minister of immigration, expressed his hope that "most provinces will agree to a program based on a Manitoba proposal to increase the number of disabled refugees admitted to Canada from Southeast Asia."[302] He informed the Commons that "a draft agreement has been worked out and submitted to the other provinces for consideration."[303] While welcoming the proposal for accepting medically inadmissible refugees when the province of residence was willing to cover the costs,[304] the federal government remained steadfast in rejecting applicants who lacked provincial financial support. On 24 February 1982, the *Star* reported that Helena Yu, the sister of a Canadian citizen born in Hong Kong, was denied entry because she was deaf and "according to immigration regulations, likely to "cause excessive demand on health and social services in this country."[305] Rev Bob Rumball, executive director of the Ontario Mission for the Deaf, criticized the decision. The reverend argued that Ms Yu, who had been educated at the Hong Kong School for the Deaf, was going to be an asset to the deaf in Toronto: "She's been getting a better education at that school than anything we have for the deaf in this country. She'll more likely provide leadership to the deaf in this country."[306] Following a review at the Immigration Appeal Board, Ms Yu was allowed into Canada when the board decided that though "the rejection was made according to the law ... there were grounds warranting an appeal."[307] A follow-up article observed: "There is little evidence that people who are deaf do, in fact, require more health or social services than the ordinary Canadians use in the normal course of their lives ... unemployment among deaf people is no greater than unemployment among the population as a whole."[308] Further indication that medical admissibility was closely tied to the perception of immigrants as economic burdens was provided in the case of Hansaraj Singh, a Guyanese married to a Canadian citizen, who had flown to Toronto in need of immediate heart surgery. Despite doctors' statements that the operation was necessary to save his life, "immigration officials refuse to extend his three-week visitor's visa, because he ... would cause excessive demands on health and social services."[309] After the story reached the press, Roger White, spokesman for the Department of Immigration, clarified that the man was not going to be deported "until after his life-saving operation,"[310] provided Mrs Singh paid the medical costs and signed a declaration to that end. Federal authorities were willing to bend the rules once guaranteed of no additional expenses.

A number of cases coming under public scrutiny in those years also indicate that people were often rejected due to disabilities that were unlikely to create excessive cost. Erring on the side of caution, immigration authorities tended to reject applicants on the basis of any condition without further investigation. Only those able to fight the rejection were later accepted through ministerial permits. The story of Helen Yu is just one example. In 1983, another immigrant woman went through similar sorts of mishaps, finally receiving permission to remain in the country. On 31 August, Joe Serge wrote an article for the *Star* chronicling Shernaz Kapadia, a polio victim from India whose parents and sister were permanent residents in Canada and who was ordered back to India because she was classified as inadmissible.[311] Ms Kapadia opposed the decision and sent letters to immigration authorities explaining: "I wear braces and crutches instead of using a wheelchair. But I climb like a monkey when there are steps to climb. I love my independence at every cost. It is true that my handicap is permanent. But my general health is perfect. I am not a burden on anybody."[312] Toronto lawyer Mendel Green defined as "outrageous"[313] the medical report suggesting the woman was likely to become a financial burden. Fortunately for Shernaz Kapadia, Immigration Minister John Roberts intervened and allowed her to stay. It is reasonable to wonder how many others were less fortunate and never given a chance because they were either ignorant of the system or their story never reached the attention of a newspaper correspondent who publicized their case.

In spite of the publicity given to several cases in the press, significant changes to the way immigration authorities dealt with medically inadmissible immigrants were not implemented in the years leading up to the early 1980s. From the articles, letters, and editorials examined, it also appears that while touched now and again by single stories of hardship, the readership and contributors to the two newspapers were reluctant to consider indiscriminate acceptance of medically inadmissible persons. As already evidenced in the analysis of parliamentary debates, one notable change over the years consisted in how rejections were justified. Whereas initially persons with disabilities were kept out because they were deemed a threat to the genetic pool of the country or were unemployable, the years after the 1950s saw the emphasis switch to the costs involved. This shift is unsurprising bearing in mind that by the end of the Second World War, eugenics had lost its appeal following the horror of the Nazi policies of racial cleansing. Furthermore, two main changes had occurred in the 1950s and 1960s. First, as seen in the previous chapter, a new attitude towards Canadians with disabilities had developed, initiated by the returns of veterans after the Second World War and their activism aimed at creating broad public support.[314] Once agreed that citizens with disabilities could no longer be deemed deviant, unemployable, and unworthy, it became difficult to advance those arguments in the case of non-citizens. Second, the Medicare debate began animating Canadian society, ending in the late 1960s when Canadians were provided with coverage of visits to hospitals and doctors as well as diagnostic services.[315] As the government grew concerned with keeping medical costs down, immigrants likely to use such services became a financial expense Canada was unwilling to incur. The concern with costs loaded onto taxpayers' shoulders reached its apex in the 1970s and 1980s when neo-liberalism, understood here as a form of governmentality centred on the concept of self-sufficient individualism, gained increasing recognition not only among politicians but also within dominant sectors of the larger society. The next chapter will further expand on this point and show that neo-liberal policies and ideas have remained strong to this day.

Another significant change was in the approach taken by contributors and readers. From the beginning of the twentieth century to the years immediately following the First World War, the *Toronto Star* and the *Globe* limited their coverage to the general policy of immigration and were supportive overall of the federal government's handling of undesirables. The tone of the discourse slowly changed in the late 1920s and early 1930s. During these years, individual cases made their appearance and while the papers remained supportive of the legislation, they began to show some sympathy for the unfortunate individuals who were penalized. Juxtaposing this chapter with the previous one, the reader will certainly note that one major difference stands out: while the stories are akin, the approach held by Parliament and the press is anything but similar. Parliamentarians were at times faced with individual stories, but their typical response was to downplay the uniqueness of each situation by appealing to the general policy. Whereas exceptions were made through the years, mainly out of convenience and because of public pressure, politicians attempted to de-emphasize the significance of these deviations by reducing to a minimum the number of cases that could constitute a precedent. Nevertheless, the press was all for giving relevance to single cases. Whether this preference was dictated by the fact that individual stories had a high selling potential or by the desire to take advantage of the charity approach to disability among the larger society,[316] the exercise helped Canadians see beyond general abstractions and understand that real lives were torn apart by the provision of medical admissibility. Undoubtedly, this was a positive improvement, albeit a limited one.

Another interesting element emerging from the analysis of newspapers' coverage is the role played by race in the exclusion of medically inadmissible immigrants. Up until 1967, the issue did barely present itself as Canada's immigration policy explicitly favoured immigration from European countries.[317] During this time, immigrants with illnesses or disabilities were unwelcome, even those coming from the mother country and even if rejections were creating tensions with Britain. After the elimination of racial discrimination following the introduction of the point system, the face of immigration significantly changed. In the period from 1971 to 2001, census data indicate that the number of people residing in Canada but born in Asia, Africa, the West Indies, or Latin America increased from 1.7 to 10.4 per cent.[318] The cases examined in this chapter illustrate how some of the immigrants who were deemed inadmissible were coming from non-traditional source countries. It is a fair guess that those newspaper readers who were uncomfortable with the idea of changing Canada's racial make-up were obviously against the admission of disabled/diseased persons coming from non-European countries. Yet, racism functions in conjunction with other social forces,[319] and this must be taken into account when assessing the role race played in affecting newspapers' approach to medical admissibility. I recognize that for some readers it might have been easier to show compassion for a white middle-class woman such as Lynn Hackett than for a racialized woman as Harjit Kumar. I am not denying that race came up in the letters of readers who, for example, were in support of deporting Mrs Kumar back to India. However, race should not be considered as disconnected from other aspects of identity. Concerns surrounding disability interacted and strengthened issues surrounding race, resulting in what Sherene Razack defines as "interlocking oppressions."[320] As shown by several scholars of Canadian immigration, racial minorities have historically been perceived as problematic, yet they have also been economically valuable assets, particularly during periods of high employment when labour was scarce. Persons with disabilities were hardly considered of any use; actually, they were barely considered at all except in terms of the burden they represented. This negative understanding of disability should not obscure the fact that, as noted by Parin Dossa, persons with disabilities, far from being all in the same position, are "hierarchically ranked."[321] As a result, whereas all persons with disabilities are devalued, it remains true that men with disabilities are considered more worthy than women with disabilities, and among women with disabilities, white women are ranked higher than racialized women.[322] Newspapers' coverage seems to validate this pattern.

While the ranked system used to assess those with disabilities persisted through time, things slightly improved in 1985 after the implementation of section 15 of the Charter of Rights and Freedoms. The inclusion of persons with mental or physical disabilities in the equality clause has forced Canadian society to reconsider its position vis-à-vis a group of individuals it had, up to that point, gladly ignored. The following chapter looks therefore at the way the Charter affected public perceptions of medically inadmissible immigrants and how some of these new beliefs were reflected in the press coverage of the medical admissibility provision.

4 Medical Admissibility: *Toronto Star* and the *Globe and Mail*, 1985–2002

The previous chapter concluded by recording no substantial changes to the letter of the medical admissibility provision in the Canadian Immigration Act in the years leading up to the early 1980s. Similarly, the coverage in the *Toronto Star* and the *Globe and Mail* underwent few – albeit in this case significant – modifications. Despite the fact that from the early 1930s some sympathy was reserved in articles, editorials, and letters to foreigners considered as medically inadmissible, overall support for a policy of exclusion remained strong. Things took a different direction in 1985 after the implementation of section 15 – also known as the equality clause – of the Charter of Rights and Freedoms. Entrenched in the constitution that was repatriated in 1982, the clause went into effect three years later "to give Ottawa and the provinces time to review their laws and eliminate any discriminatory provisions that could not be defended in court as reasonable limits."[1] Section 15 provided a means to challenge many of the laws and statutes that were perceived as discriminatory by a number of identity groups, including persons with disabilities. In this chapter, I discuss the impact of the Charter on medically inadmissible applicants as reflected in the coverage provided by the *Globe and Mail* and the *Toronto Star*. As in the previous chapter, my goal is not to identify the ideological positions of various associations whose comments appeared in articles, letters, or editorials. Rather, I want to convey the general tone of the public discourse formulated in the press. The purpose is to demonstrate that through time the equality clause has affected the press's approach to medical admissibility. As Freedman argues, while the press has acquired substantial independence and "is by no means the direct expression of a state's political priorities, it makes little sense to ignore the impact of political actors and political values on the character of the wider media environment."[2] The impact of the Charter was indeed significant. The legislation was not reversed or substantially modified, yet the Charter highlighted the concerns of those excluded on medical grounds and, as I will discuss in the next chapter, provided them with a powerful tool to contest the policy in a court of law. With respect to press coverage, it contributed to a more sophisticated analysis of the medical admissibility provision. The deeper evaluation does not necessarily mean that the press became critical of the provision, but it would be impossible to deny that, confronted with the rights discourse at the core of the Charter, newspapers were forced to revise their position of unquestioned support and provide a more nuanced and articulated rationale for maintaining the provision.

Adjusting to the New Reality of the Charter, 1985–1990

The first outcome of the Charter was felt within days, when on 31 January 1985, the Conservative government introduced Bill C-27 to amend various laws that were "in clear violation of the Charter of Rights and Freedoms."[3] Among others, the bill eliminated "statutory powers of inspectors ... to enter private homes without the consent of the owner or without a judicial warrant," "the power of immigration tribunals to hold in camera hearings," as well as the power of the Human Rights Commission "to select members of tribunals that rule on cases of alleged discrimination raised by the commission itself."[4] Change was slow, with Liberal and New Democratic Party critics accusing the government of dealing "only with the easy and non-controversial issues."[5] Indeed, many of the more difficult legal questions – such as those "involving the equality rights of the sexes, racial and ethnic minorities, and the handicapped"[6] – were altogether ignored. According to Dale Gibson, the legislature was reluctant to act decisively in changing previous legislation since it was unclear "how generously the courts would construe the equality guarantee."[7] Consequently, the courts were soon presented with a number of issues that had gone unresolved at the legislative level. On 15 August 1985, four months after the implementation of section 15, the *Globe and Mail* reported on the case of a mentally disabled Indian citizen who had been refused immigrant status; acting on her behalf, the Canadian Association for the Mentally Retarded[8] challenged the decision and the immigration law as unconstitutional because it was in violation of the Charter.[9] Orville Endicott, the association's legal counsel, explained, "the woman's right under the Charter to equal benefit of the law is being violated because she is being discriminated against on the basis of mental disability. Mental or physical disability is a prohibited ground of discrimination under the equality provisions of the Charter."[10] While the paper did not follow through with the case, it is likely the court dismissed it as the issue of whether the Charter applied to non-Canadians had not yet been resolved.[11]

As noted in previous chapters, Canadian immigration authorities abroad tended to exclude all applicants with disabilities, without considering whether the condition was going to have a negative impact on the person's economic contribution to society (all other possible contributions were hardly appraised). Furthermore, officials abroad were often poorly equipped to conduct reliable tests, thus mistakenly rejecting persons who were not affected by any disease or disability. On 4 October 1985, the *Toronto Star* reported that Abrar Zuberi, an immigrant from Pakistan, was denied entry because he was diagnosed with tuberculosis. After a series of tests in the United States, Mr Zuberi was however declared in good health. Being a British national, he was able to reach Toronto and there as well doctors concluded he was a healthy man. Commenting on his ordeal, Mr Zuberi declared that he hoped "something is done about the way medical reports are handled in India. I was lucky. I was able to come to Canada and speak out. Others are left high and dry."[12] The case came under scrutiny by the immigration section of the Canadian Bar Association; Toronto lawyer Carter Hoppe, vice-chairman of the association, expressed his concern that "in increasing number of cases, parents, spouses and children are denied sponsorship for medical reasons – although independent medical examinations show them to be healthy."[13] Evidently, Health and Welfare Canada doctors overseas "tend to err on the side of caution."[14] This guardedness resulted in doctors becoming "paranoid" in front of any indication of disease or disability "in the family medical history"[15] of the applicant.

Another story publicized in the press concerned Canadian veteran and former prisoner of war (POW) Fred Darvin, a sixty-nine-year-old who was born in Manitoba and had later become a US citizen. In 1985 he applied to come back to Canada where his family resided. Mr Darvin was deemed inadmissible because of his health condition. Since 1972, the man had "lost a kidney, part of his stomach and had his left leg amputated when gangrene developed."[16] As soon as the news appeared in the paper, both provincial and national leaders of the Royal Canadian Legion[17] mobilized and the Department of Employment and Immigration came under attack. Discarding the department's argument that "Mr. Darvin gave up his Canadian citizenship and is no different than any other U.S. immigrant,"[18] several Canadians were outraged by the decision of denying entrance to someone who had risked his life for Canada. Responding to the department's statement that Mr Darvin was likely to represent an economic burden, Ed Slater, director of service for the Dominion Command of the Royal Canadian Legion in Ottawa, dismissed it as nonsense. Fred Darvin was not going to be a drain on health services since he was entitled to "certain rights because he served his country during the Second World War. He's a veteran."[19] Mr Slater added, "This man fought for his country and he was a POW [prisoner of war], and if he was good enough to fight for Canada, then we should welcome him back."[20]

Considering the flood of support Mr Darvin received from numerous sectors of Canadian society, Employment and Immigration Minister Flora MacDonald agreed to review the case and granted the man permission to move back to Canada. In her message to the press, Mrs MacDonald declared that "she was only made aware of the case through The Globe story"[21] and that she was willing to issue a ministerial permit to Mr Darvin. The minister's explanation appeared to have little credibility, especially after the *Globe and Mail* reported that Veteran Affairs Minister George Hees had written her twice during the previous months pleading for reconsideration of the case.[22] Whether or not Mrs MacDonald was telling the truth is irrelevant to my argument. More significant is the fact that Fred Darvin's story brought forward deeper questions surrounding the value of citizenship. Canadians were faced with the problem of applying immigration regulations to someone they considered one of their own, creating a contentious battlefield between government authorities and certain segments of civil society. The latter perceived Fred Darvin as Canadian, while the federal government regarded him as a foreigner who had renounced Canadian citizenship, thus giving up the rights associated with it. The government made a considerable mistake by undervaluing the impact that issues concerning veterans have always had on public opinion. Whereas Canadians were willing to follow the government in its neo-liberal approach to "social policy along lines of economic participation"[23] rather than membership in the national community, the deal broke down when applied to people who had fought and risked their lives for the country. War, veterans, and the rhetoric surrounding patriotism have always had a highly emotional impact on civil society.

The mid-1980s also witnessed the emergence of new concerns regarding immigrants with HIV or AIDS. As usual, the US government took the leading role. On 4 February 1986, the *Toronto Star* reported that US secretary for the health and human services, Mr Otis Bowen, had signed an order "requiring all permanent residents and immigrants to the United States to be tested for exposure to AIDS."[24] Once approved by the White House, the order was going to add AIDS to the already existing medical conditions barring immigrants from the country. Validating the common belief that AIDS was a disease affecting those who were outside the boundary of society, the paper remarked, "The disease is most likely to strike homosexuals, abusers of injectable drugs and hemophiliacs."[25] It suggested that the people affected were burdens since "about half the more than 400 cases of AIDS diagnosed in Canada have died."[26] In *AIDS and the National Body*, Thomas Yingling writes that persons affected by HIV or AIDS have long been considered outside the limit of the ideal national body; if this notion holds true for citizens, it becomes unshakable in the case of those who are non-citizens and already considered outsiders.[27] In fact, foreigners with a disease or disability are constantly represented in the common imagery as undeserving others whose only purpose in migrating to another country is taking advantage of benefits and entitlements that, by definition, do not belong to them.

The debate around the best course of action when confronted with applicants with HIV or AIDS soon reached Canadian shores, though Ottawa initially refused to follow the example of its southern neighbour. In March 1986, the health department stated it had "no plans to test all would-be immigrants for AIDS antibodies."[28] The decision did not signal disagreement with the US policy, but was made "because of a lack of testing facilities and because it is possible to get false results."[29] A report prepared by the Canadian Bar Association asked for all immigrant applicants to be tested, though Dr Alastair Clayton, director of the federal government's Laboratory Centre for Disease Control, rejected the suggestion as "premature because the test available is inconclusive" and it "is not available in all countries."[30] Even if disregarded, the request put forward by the association indicates that sectors of society were firm in their opposition to the entrance of persons with the disease and that several government representatives were not prejudicially against immigrant testing. In its report, the association stated, "immigration to Canada is a privilege and not a right. Individuals who show a positive test of the AIDS antibody have a 5 to 30 per cent chance of developing AIDS ... and those immigrants are expected to contribute to Canadian society."[31] Voyvodic observes that the belief that immigrants have privileges but not rights remains the major hurdle in their fight for integration within Canadian society.[32] Privileges can be easily conferred and withdrawn. Having rights is certainly no assurance of life, liberty, or equality (too many examples throughout history demonstrate it), but it nevertheless represents a reassurance that the individual in question belongs to a community.[33] That belonging is the precondition for claiming any other benefit. In *Exalted Subjects*, Sunera Thobani illustrates how these rights, defined as "the nation's privileges,"[34] operate to simultaneously discipline insiders and exclude outsiders. Whereas we cannot deny the unbalanced distribution of these rights among the national population (men above women, rich above poor, healthy above diseased, able-bodied above disabled), we can certainly agree that the passive acceptance of this unbalance among the lowest sectors of the citizenry rests on the guarantee that, no matter how low some national subjects are positioned within the community, they are still part of it. This hierarchical structure is made evident in the comparison with outsiders who, lacking access to the community, cannot ever claim the nation's rights to themselves and are, therefore, excluded a priori from the enjoyment of certain entitlements.

The debate continued in subsequent months as the United States followed through with its plan for AIDS testing of immigrants. On 9 June 1987, the *Toronto Star* reported that "only a week after President Ronald Reagan called for "routine" AIDS tests for immigrants, prison inmates and people attending sexual-disease clinics,"[35] Attorney-General Edwin Meese announced the creation of a new "massive AIDS testing program" whose goal was to "bar all 'immigrants, refugees and legalization applicants' who test positive for AIDS."[36] The news sparked further discussion in Canada. On 10 June 1987, the *Globe and Mail* published a statement by Scott Leslie, chief of the department's immigration and quarantine section, explaining that Canada operated differently from its southern neighbour and did not conduct tests for AIDS. While the United States had decided to ban all immigrants and refugees who tested positive for AIDS antibodies, Dr Leslie clarified that "people who test positive to AIDS antibodies will not be prevented from immigrating to Canada."[37] Despite reassurances to the contrary, Ottawa was quick to change its mind about testing immigrants. Three months later, the *Globe* informed the public that Health Minister Jake Epp confirmed that "the federal Government is reconsidering its decision not to make AIDS tests mandatory for prospective immigrants."[38] Giving "no indication of how soon a decision would be made," Epp told the House of Commons that "the federal Immigration Medical Review Board, an advisory group for the departments of Health and Immigration, has reversed its position and now recommends AIDS testing of immigrants."[39] The minister added that "his department is under pressure from some public and professional groups to introduce mandatory testing of immigrants."[40] The paper was careful not to present the suggestion as a fait accompli. The article included comments by Dr Alastair Clayton, director general of the federal Laboratory Centre for Disease Control, who downplayed the possibility of imminent policy change. According to Dr Clayton, "Mandatory testing of immigrants offers false impressions of improved security for Canadians and creates a nightmare of procedural difficulties."[41] Additionally, there was the question of "what to do with immigrant applicants and refugees already in Canada who test positive for AIDS exposure."[42]

In November, Dr Clayton reiterated his opposition to mandatory testing, arguing that it was not an effective solution since, even if testing was conducted, it would only identify a number of people "so small the resources would be better spent on education."[43] Other experts disagreed. Tracey Tremayne-Lloyd, national health law chairman for the Canadian Bar Association, defended the idea as a means to spare Canada "enormous expense just in terms of the medical costs."[44] Dr Norbert Gilmore, chairman of the National Advisory Committee on AIDS[45] warned, "it currently costs $80,000 to $100,000 in medical expenses alone to care for a person with AIDS."[46] Ms Tremayne-Lloyd emphasized, "the obtaining of immigration status is a privilege and not a right."[47] Other experts suggested that the money should be spent on education rather than testing. Bob Tivey, executive director of the Stop AIDS Project in Vancouver, noted that the number of those infected in Canada and the United States was getting so large that only naive people could seriously believe "immigrants' testing would prevent the illness from spreading."[48] As usual, immigrants became a scapegoat for a problem whose magnitude seemed unprecedented.

The debate was far from over. On 6 June 1989, the *Globe and Mail* reported that researchers at Montreal General Hospital and Queen's University in Kingston concluded, "screening could have prevented 310 to 780 HIV-infected immigrants from entering Canada."[49] Speaking at the Fifth International Conference on AIDS held in Montreal, Dr Hanna Zowall, epidemiologist and one of the researchers at the Montreal General Hospital, explained that in a ten-year period, "screening could save between $5-million and $17.1-million in hospital costs."[50] At the same time, researchers acknowledged that they "did not take into account the social, legal or ethical implications of screening potential immigrants for AIDS."[51] They also conceded that tests were not 100 per cent accurate and "the suffering of people who are falsely identified as positives has to come into the equation as well."[52] Critics in the audience were sceptical of the claim that screening could save money. Jesse Green, researcher at New York University Medical Center commented, "The cost of treating AIDS patients in hospitals has been grossly overestimated. AIDS is cheaper to treat than previously believed."[53] Green's observation was factually accurate but, as we have already witnessed, facts have a small role to play when relying on people's fears is so much more convenient.

AIDS was only one of the issues facing the immigration department and getting public attention during the second half of the 1980s. On 3 March 1987, the *Star* recounted the ordeal of an Oakville couple, both of them Canadian citizens, who were refused permission to bring relatives to Canada because one of them was diagnosed as mentally retarded and had cerebral palsy. The case is interesting because it shows how immigration rulings can be successful in pitting immigrant applicants against each other, thus taking advantage of divide-and-rule tactics, whereby the oppressed focus on fighting each other rather than opposing the source of their oppression. In this case, the couple accused the immigration authorities of discriminatory behaviour after learning that "a Soviet cancer victim was recently allowed to immigrate to Canada with her husband and daughter for medical treatment."[54] They argued that the cases were similar and should have received the same consideration. Frank Rodriguez, immigration appeal board officer, rejected the claim as "there are insufficient grounds to admit the family for humanitarian reasons because there is no evidence of 'undue hardship,' as in the case of the Soviet family."[55] While emotionally understandable, the frustration of some applicants was misdirected. Systems of oppression thrive on the conflicts arising among equally powerless subjects since these clashes elude questioning the real source of oppression.

With immigrants scrambling for acceptance, it is worth assessing Canadians' prevalent perceptions with respect to immigration. Throughout the 1980s, political parties were wary of alienating the increasingly powerful ethnic vote, especially in major metropolitan centres. At the same time, however, they were aware that different segments of society opposed Canada's pluralistic immigration policy fearing ethnic unbalance and the economic impact of immigration, particularly during periods of unemployment.[56] On 17 September 1987, an advertisement appeared in the main section of the *Globe and Mail* presenting the results of a national poll on people's opinions around the current state of immigration policy. Similar full- and half-page ads were placed in several other newspapers across the country. According to their content, Gallup Canada had been commissioned to conduct the poll by the Immigration Association of Canada, "a privately funded non-profit and non-partisan agency that grew out of a general concern about immigration policy."[57] One thousand and forty-eight people were interviewed. The results indicated that 69.6 per cent of the interviewees believed that "Canada's immigration policy should be designed in the interests of the majority of Canadians and that the entry of immigrant workers and their families should be geared to Canada's social, cultural, and economic needs, without changing the ethnic and cultural composition of the country."[58] Only 30.4 per cent of the people contacted agreed with the statement that "Canada's immigration policy should be designed to provide a safe haven in Canada for the disadvantaged people of the world who may be fleeing from political oppression or poverty."[59] On issues of health, the questionnaire asked: "Should immigrants preparing to enter Canada be satisfactorily cleared for all aspects of health, as well as criminal and subversive behaviour, before being permitted to enter, without discretionary power being left to the politicians?"[60] The results were almost unanimous, with 94 per cent answering YES and 6 per cent answering NO. Unsurprisingly, "the ads triggered an avalanche of criticism from individuals and organizations across the country."[61] While the immigration department refused to comment, critics accused the poll of being "blatantly racist and anti-immigration."[62] Much of the criticism was over the language adopted in the questionnaire. Despite the claim of giving "a fair and accurate reflection of Canadian opinion,"[63] the questions were formulated in such a way that the end result was a general impression of unfriendliness towards immigrants. The first question asked people to choose an immigration philosophy at the expense of the other: Was Canada promoting immigration for its own interests or out of generosity? No middle ground was offered. The sixth question asked whether immigrant applicants should have been cleared "for all aspects of health, as well as criminal and subversive behaviour, before being permitted to enter."[64] Combining health and criminality in the same question implicitly suggests that the two are akin and should be similarly dealt with. Polls might appear objective to the inexperienced eye, but they channel the interviewees' answers in precise directions that often leave little space for nuances.

Whether or not the sample in the poll reflected public opinion, it seems that most Canadians were unwilling to accept similar treatment towards their nationals abroad. On 22 October 1987, the *Globe and Mail* recounted the story of Patrick Worth, a Canadian citizen from Toronto, who had been detained at the airport by US immigration officials trying to determine if he was mentally retarded. John Ingham, deputy district director of the US Immigration and Naturalization Service, explained, "Under U.S. law, a mentally retarded person is not allowed to enter the country alone."[65] The detention prompted Mr Worth to ask Ottawa to "take up the issue with Washington."[66] He found the conduct of US authorities offensive and discriminatory towards persons with disabilities: "Why can't I, just as a Canadian citizen, be able to travel into the United States without being harassed?"[67] The paper sympathized with Mr Worth; however, it correctly pointed out that equality and fairness seemed to work one way only, namely when Canadians were involved. The article commented that readers were comfortable complaining about the discriminatory treatment aimed at a Canadian citizen by US officials, but they gave little attention to the equally discriminatory approach that Canadian immigration authorities had recently shown when an American family headed to Alberta was refused entry because the son had cerebral palsy. Evidently, the right to respect and fairness was easier to assert when Canadians were the beneficiaries.

The two previous cases demonstrate that several Canadians were still uncomfortable with disability. In this context, David Suzuki's article, published in early 1988, for the *Globe and Mail* about mankind's "unprecedented ability to alter the genetic makeup of life forms"[68] bears mentioning. Going over the history of genetics and its developments, Suzuki, a Canadian world-renowned geneticist, academic, broadcaster, and activist, argued that among its consequences were also tragic outcomes in several countries around the globe. In the United States, for example, "direct results of this enthusiasm ... were the imposition of immigration restrictions against people considered 'inferior,' the sterilization of mental patients and the prohibition in many states of interracial marriages."[69] In Germany, this downward spiral had reached its apogee when "doctors and scientists ... concluded that human beings could be 'perfected' through selective breeding and the elimination of 'defectives.' The Nazi Race Purification programs seemed to represent the application of some of the most 'progressive' ideas in science ... doctors and scientists of that earlier time – especially geneticists – had ... popularized the notion of the overriding importance of heredity in human behavior and sold it to Hitler's National Socialists. It led inexorably to the horrors of the Holocaust for which scientists must therefore acknowledge some responsibility."[70] Whereas some of his colleagues refused to take the blame and accused Suzuki of being "hysterical,"[71] he kept warning that "this selective memory of science's history amounts to a coverup and a revisionism that only ensures that it could happen again."[72] Indeed, rejection of disability has persistently continued to frame the public discourse in almost every sector of society, from the scientific community to government officials.

The current flourishing of private eugenics means that unborn children diagnosed with chronic diseases and disabilities can be, and often are, aborted under the excuse that theirs would be a life not worth living. As Bröckling remarks, "One may fully support the right of a mother to decide to abort an embryo with Down's syndrome, but there can be no doubt that this decision is an individual eugenic choice."[73] As documented by Liz Crow, public discourse is increasingly tolerant of practices such as euthanasia and infanticide in those cases where life reaches a quality so low that it is considered no longer acceptable.[74] In current public discourse and perception, certain lives are assessed as unvaluable and therefore unworthy of existence, at both the individual and societal levels. I argue that these assumptions about unworthiness must be examined and understood within the current political, economic, and ideological context of extreme individualism. The neo-liberal belief that individuals are exclusively responsible for their fate has resulted in the celebration of independence and the perception that any form of community help or support is a failure. As a consequence, the moment people become incapacitated and cannot adequately provide for themselves, they get branded as burdensome to others and their life gets automatically reassessed as low quality and unvaluable.

While segments of society continued to believe they had the right to decide which lives were worth living, those considered less-than-normal began fighting back and arguing that their rights should be recognized. The movement gradually spread to the larger society where some people became more conscious about disability and less willing to accept blatant discrimination from their government against physically and mentally disabled individuals. Faced with immigrants' complaints and public criticism, Immigration Canada also came under scrutiny from the judicial system in those years. On 8 March 1987, the department filed an appeal against the order of the Canadian Human Rights Commission tribunal to compensate Mehran Anvari, an Iranian citizen who was denied landed immigrant status "because of his disability, a consequence of polio."[75] Despite the department's argument that Mr Anvari's "leg problems and scoliosis ... could prove costly to Canada's health and welfare system,"[76] the tribunal found that the rejection had resulted in "injury to feelings and self-respect"[77] and was contrary to the Canadian Human Rights Act passed in 1985. In a letter published in the *Star* on 16 March 1989, in response to the ruling, a reader from Toronto, Kazik Jedrzejckaz, expressed his support of the tribunal's decision and criticized immigration officials: "It is disgraceful that the Immigration Department is appealing this decision, wasting the taxpayers' money ... They are afraid that Anvari, stricken with polio, will be too costly to Canada's health and welfare system. But surprisingly, our government is not afraid to spend $9 billion on nuclear-powered submarines. My face is red from the shame."[78] The government's rationale for austerity and cost-saving measures systematically fails to mention that while some spending is considered superfluous and must therefore be eliminated, money is allowed to flow freely and without question on other selected budget lines.

The debate surrounding medically inadmissible persons reached new heights when the story of Miguel Silva, a nine-year-old boy born in Peru but living with his family in Toronto, received public attention after an article was published in the *Star*. The boy was denied landed immigrant status because he was born with Down's syndrome. His parents had sought refugee status for the family, yet Miguel was ordered back to Peru. Criticizing the department, the paper concluded, "Miguel's family has had no choice but to launch a costly court challenge under the Charter of Rights. That's absurd."[79] Some Canadians were ashamed, as evident in the letter sent by Mary Bennett of Pickering. Considering the effects such decisions were going to have on persons with disabilities in Canada, Ms Bennett wrote, "This is a giant step backward for our handicapped.

Over the last few years the push has been to get our handicapped out into the community and yet here we are saying no there is something chromosomally wrong with you, we don't want you in our country."[80] Similar outrage transpired in the letter of Bonnie McDowell: "Our Conservative government denies landed immigrant status to a Peruvian family who do not have a criminal past but do have a 9-year-old son with Down syndrome. The government, ever mindful of our tax dollars, feels that this young boy will be a burden to the Canadian taxpayer. Thank you, Prime Minister Brian Mulroney and Immigration Canada for your concern, but I feel the burden on my tax dollar is having to support a group of inept, insensitive, self-serving bureaucrats."[81] Commenting on the discrimination embedded in the country's immigration laws, Orville Endicott, lawyer for the Canadian Association for Community Living, remarked, "I don't think the average person has any awareness of how a disability figures in people's chances to immigrate."[82] Diane Richler, vice-president of the association, "told reporters the laws assume the worst from the start and, as applied, have become unduly restrictive."[83] Asked a few days later to comment on the case of another child denied landed immigrant status because they had Down's syndrome, Ms. Richler added: "These situations are sending a message to all Canadians who have a mental or other handicap about their value as citizens ... it's very reinforcing of the image of people with handicaps as having no value. You're telling people that if you had the choice, this country doesn't really want them."[84] The *Toronto Star* called for changes to the immigration law. Referring to the two cases that had recently occupied the front pages of the paper, it pondered: "What kind of country would split up a family by granting landed immigrant status to its healthy members while turning away a child with disabilities? Regrettably, Canada, for one."[85] The decisions also contradicted what an official of the department had told the press back in 1958[86] – namely that the official policy was to reject "the entire family rather than permit the family to be split up."[87]

Ms Richler's comments to the *Star* reveal the interrelation between immigration policy and citizenship. The former is understood as a means of building and developing the national body and, as a consequence, some applicants are deemed more valuable than others. Whereas rejection of persons with disabilities would be unthinkable when dealing with citizens, it is not merely acceptable but actually encouraged in the case of foreigners. Yet, if we limit our focus to disability only, no one has provided a sound rationale explaining why a person with a disability should be considered a burden if coming from outside the national boundaries yet is a worthy member of the community if born in the country. In order to understand why such an apparent inconsistency is allowable, we need to reframe the discourse by looking at disability as an aspect of identity that cannot be separated from race and ethnicity in the daily experience of real individuals who are also members of specific communities. In this sense, we need to reconsider the issue in light of Thobani's argument that while nationals can and are differently positioned depending on their social identity, they are by definition included in a group that enjoys certain benefits and entitlements exclusively on the basis of nationality.[88] Nationals with a disability cannot be assessed the same way as outsiders with a disability because they are inherently separated from those outsiders. Disability as an identity category cannot be considered distinct from other identity categories.

Two weeks later, an editorial provided a brief account of gains and unsolved problems that the disabled community faced in Canada. Reflecting on the treatment received by persons with disabilities in the country, the editorial recalled, "For decades, disabled people in Canada were either locked away in institutions or held virtual prisoners at home, well away from an intolerant public ... Just eight years ago, disabled rights activist Beryl Potter was told she'd have to buy two plane tickets if she wanted to fly to England because nobody would want to sit next to someone who was missing three limbs ... some attitudes ... remain mired in the ignorance of the past."[89] Noticing that discrimination remained a reality in the lives of persons with disabilities, the writer observed that different sectors of society were doing little to change the status quo. Among others: "Government is not blameless. Ottawa is also guilty of perpetuating discrimination. For example, Immigration Canada refuses to grant landed immigrant status to some children with Down syndrome ... As a society, Canada has encouraged the disabled to come out of hiding. But, clearly, there is much more to do."[90] The message seemed to resonate when Ottawa decided to review the section of the Immigration Act preventing admission of persons with disabilities. Revealing that the Japanese girl with Down's syndrome and her family were being allowed to stay, the Conservative MP for Don Valley North, Barbara Greene, acknowledged, "This provision of the act is just incompatible with most people's thinking. The presumptions about the handicapped in the act (are that) they will be a burden on society, which – with the proper help and support – they won't be."[91] Rita Mezzanotte, spokesperson for Employment and Immigration Minister McDougall, confirmed that officials in the immigration and health departments were entering into discussions on how to ease the review process for persons with disabilities so that it could abide by "the provisions of the Charter of Rights."[92] Commending the government for its actions, the *Star* published an excerpt from an editorial first printed in the *Edmonton Journal* on 25 July. The editorial defined as "encouraging ... that the government is considering changes to provisions of the Immigration Act which discriminate against disabled people."[93] The move was appreciated considering: "Many disabled Canadians have managed to lead productive, useful lives, contributing as fully to society as able-bodied citizens. They would rightfully resent any assumption that their condition automatically places a strain on the country's system of health care and social services."[94] Good intentions aside, concrete measures were not forthcoming. This shortfall prompted the Canadian Bar Association to declare that "Canada's immigration rules are so outdated that officials are rejecting people with part of a finger missing or who suffer from minor skin conditions or high blood pressure."[95] Individuals were denied admission because authorities felt they could constitute a burden for the health care system. However, according to Winnipeg lawyer Mira Thow, "there isn't any list immigration officers can use to check whether specific services are in short supply."[96] The federal government set up a system to solve this small problem and came up with its own list in the record time of less than five years.

Asking the Courts for a Chance to Stay: 1991–2002

The early 1990s witnessed increased frustration among applicants who were denied admission on medical grounds. Unfortunately, this exasperation resulted in applicants blaming each other rather than focusing their combined energies in fighting a law that was discriminatory towards all of them. Applicants expressed their dissatisfaction when Ottawa opened its doors to refugees while keeping out those who, despite a disability, had enough capital for self-support and met the other immigration requirements. On 10 May 1991, the *Globe and Mail* published an article about a former Canadian living in Southern Ontario who was prevented from regaining her citizenship because she had cancer. Diane Smith, a nurse's aide, was born in England in 1944, the daughter of "a British war bride who served in the Royal Canadian Air Force and a soldier with the Royal Canadian Army Medical Corps."[97] While in England, she was unaware of legislation passed in Ottawa in 1947 requiring natural-born Canadians to register if they intended to retain their citizenship and had therefore lost her status. Upon returning to Canada, she was denied landed immigrant status because she had lung cancer. The immigration department claimed, "we are complying totally by law,"[98] yet Diane Smith remained of the opinion that she was entitled to Canadian citizenship. Mrs Smith did not blame Ottawa for the intrinsic unfairness of the decision; rather, she felt she was denied her status because "Canada is opening its arms to many people with no ties to the country, and ... it bends over backward to accommodate people claiming refugee status who are convicted criminals or wanted for crimes elsewhere."[99] Those seeking entrance into Canada fail to realize they are in the same boat and often turn against each other rather than unite in fighting the legislation. Their inability to create a coalition opposing the legislation and presenting a different narrative to Canadian society allows Ottawa's discourse to go unchallenged. The lack of an alternative narrative is all the more damaging when we consider that uneasiness towards refugees, people convicted of crimes, and immigrants in general is widespread among many Canadians. Throughout the first half of the 1990s, demands for improved security, better screening at the border, tougher deportation procedures, and reduction in immigration quotas were common.[100] Such requests were answered in the new immigration guidelines that Jean Chrétien's Liberal government adopted on 3 November 1994. The guidelines established a lower annual immigration quota, required those accepted to assume a greater share of their expenses for settlement, set up bonds for family-sponsored immigrants, and gave more relevance to newcomers' knowledge of English or French.[101] During the 1990s, despite the rhetoric around multiculturalism and its opportunities, immigration remained a tool meant to benefit Canada economically while only minimally changing its racial and ethnic character.

In 1991, another story caught the press's attention. A Norwegian family, the Bakkeskaugs, had moved to Canada the year before and bought a farm in British Columbia. At the time of their application in 1989, Mr and Mrs Bakkeskaug were informed that one of their children, Kjetil, had Hodgkin's disease[102] and had to undergo medical treatment. The doctor conducting the tests for the immigration department assured the parents that the disease was curable and, once the treatment was completed, the son could rejoin the rest of the family in Canada. Confident in the doctor's judgment, the family moved to Canada while waiting for Kjetil to end his chemotherapy in Norway. Several months later, after successful completion of the treatment, Kjetil was ruled medically inadmissible. According to the Immigration Act, "the failure of any dependent children to qualify is sufficient grounds to refuse the family application."[103] As a consequence (and even though, as already noted, the provision was applied inconsistently over the years), the family was ordered back to Norway. In an attempt to appeal the decision, the family's lawyer took the case to the Federal Court of Canada, but Justice Paul Rouleau dismissed the appeal. Judge Rouleau concluded that the doctor's words were "gratuitous comments holding out hope and I cannot accept that [the doctor] had any authority to make any representations or promises on behalf of the department."[104] In other words, too bad if the Bakkeskaugs had been naive enough to trust someone who did not have the power to make a final decision on admissibility to the country.

Once the news reached the public, letters in support of the family began to pour in at the *Globe*. On 9 October, G.D. Elkin of Willowdale, Ontario, wrote, "to the rest of the world, the Canadian Immigration Department must [*sic*] regarded as something out of *Alice in Wonderland* where lunacy reigns supreme."[105] Commenting on the decision to deport the entire family, the reader fell victim to the same logic Diane Smith had embraced a few months earlier: "Contrast this with the criminals, drug runners and murderers who have been allowed into the country under the guise of being refugees, and you are forced to conclude that the Immigration Department is modeled after Orwell's *Nineteen Eighty-Four*, where bad is good, evil is encouraged, thrift and hard work are chastised, lies are rewarded and truth is punished."[106] In an interview given in 2007 to *Upping the Anti*, a radical journal of theory and action, Dan Irving remarks that the state thrives on the tensions arising among minority groups competing for recognition and rights. Although focusing on a different minority group, Irving's work helps in understanding how the capitalist state operates to keep apart subsections of the population in order to weaken their opposition to its policies. The trans activist and teacher explains that "within a liberal democratic context, we are led to believe that – as segmented groups representing specific atomized interests – we are in competition with other groups for rights, legal protections, access to health care, education, and essential services."[107] This competition hampers efforts towards solidarity and prevents any resistance against the state and its institutions. It results in a dismal failure of the oppressed and a victory for those who aim at maintaining the status quo. Unless a new narrative is presented, the dominant discourse cannot be truly challenged, let alone dismantled.

Different was the message contained in another letter to the paper published on 10 October. Recalling his family's experience, Alfons Mueller told the story of his Canadian-born son who, after the family moved to Switzerland, had also been diagnosed with Hodgkin's disease. The Muellers were permitted to stay in the new country with their child who "underwent chemotherapy treatment and now, 12 years later, he is completely recovered."[108] Appealing to the Canadian immigration authorities to accept the Bakkeskaug child, Mr Mueller brought forward the example of his son who "turned out a responsible and caring human being and will probably produce much more, over his lifetime, than what the treatment costs were at that time. We consider that our costs turned out to be a solid investment in a human life."[109] A letter sent by Sharon Edmundson of Selkirk, Manitoba, was instead more self-oriented. While conscious that "conditions that place a demand on our already strained health care system cannot be overlooked,"[110] the reader came forward to "propose a simple solution. Waive the exclusion order for the family members already here. The son in Norway can apply independently next year when he turns 21. His health status can be re-evaluated at that time. This solution would not tax our health services and we, as Canadians, would continue to benefit from the family's contribution to our economy and society."[111] Sure a simple solution, though a little odd for a country whose immigration policy claims to foster family reunification. Fortunately for the Bakkeskaugs, the government of British Columbia intervened and "raised no objection to the medical condition of Kjetil."[112] Reassured of the province's willingness to receive the young man, "the federal government reversed its decision and a minister's permit was issued."[113] The three letters mentioned above, though written from different perspectives, reveal a basic understanding that immigrants should be assessed exclusively based on their usefulness to the country. Within a neo-liberal capitalist system, people are explicitly considered to be producers and consumers. Society looks at them as contributors. Within such a restricted social discourse, disabled persons are systematically presented as "those who possess a faulty body, mind, or sense, which puts the ability to work at risk."[114] As Barnes and Mercer observe in their study on the interlinkages between disability and work, today, as in the past, "paid employment is central to social inclusion."[115] The assumed inability of individuals with disabilities to contribute economically to society is enough to relegate them to the status of unworthy subjects who must be excluded.

The Bakkeskaug's case highlighted some of the inadequacies of the legislation regulating immigration of families to Canada. In particular, the rules governing the admission of children with medical conditions were susceptible to criticism. In this context, in November 1991, Ottawa announced that new rules were going to take effect; immigrant children over the age of nineteen would be no longer automatically allowed into Canada with their parents, "unless they are full-time college students or too disabled to care for themselves."[116] According to Immigration Minister Bernard Valcourt, the restriction was necessary to halt the immigration of large numbers of people who "automatically qualify for visa, regardless of their job skills."[117] The new legislation appeared at odds with the current law denying immigration to "people with mental or physical disabilities that might make them a burden on the health care system."[118] Various immigration experts noted, "The reform also sets the scene for bureaucratic battles because a visa officer may use disabilities to justify issuing a visa while health and welfare department officials can then deny the visa for medical reasons."[119] Despite the immigration spokesperson's reassurances that contradictions were going to be solved before the new law took effect, the move appeared ill-conceived.

The early 1990s brought to the forefront again the issue of immigrants' testing. In November 1992, the *Globe and Mail* reported on the conclusions of a review panel led by Neil Heywood, assistant director of policy, planning, and education for Health and Welfare Canada, and Brian Grant, director of control and enforcement policy for Immigration Canada. The review team asked Ottawa to eliminate the test for syphilis required of all immigrant applicants. Considering that the test had been introduced more than forty years earlier, Dr Heywood observed, "syphilis was considered a potential public health risk at the time ... But with the evolution of medicine, there's a need to bring immigration policy into line."[120] As for AIDS testing, the panel advised Canada to "not follow the lead of countries such as the United States, France, Australia and China," all of which required "all long-term visitors and immigrants to take a blood test showing they do not have AIDS or HIV."[121] The recommendation took into consideration the fact that even those who had initially called for AIDS screening of immigrants were having second thoughts. Walter Schlech, HIV researcher at Dalhousie University, noted: "We've learned more and more about the disease. We now know that, if an individual is simply HIV positive, it could be 10 to 15 years before he needs treatment for AIDS. There are many more factors to consider than there used to be. AIDS is not a danger to public health because, unless you're talking about being raped, it takes two to tango."[122] Nonetheless, the review panel maintained that while the screening should not have been mandatory for all immigrants, those "who already show clinical signs of having AIDS will still be asked to take the test. Those who test positive will not be admitted."[123]

Several researchers dissented with the recommendation of excluding applicants with AIDS, as they were not convinced that the cost of treating those with the disease was significant. Based on a recent study comparing the cost of treating immigrants with HIV and AIDS to the cost of treating those with coronary heart disease, researchers estimated that "484 of the 161,929 immigrants who entered Canada in 1988 were infected with the human immunodeficiency virus and treating their HIV-related illnesses would cost $18.5-million over ten years ... In comparison ... 2,558 of the immigrants would develop heart disease, at a cost of $21.6-million over the same period."[124] If mandatory AIDS testing was to be implemented, so should testing for heart disease. Any other solution "would be arbitrary at best and discriminatory at worst."[125] John Blatherwick, chief medical officer of health in Vancouver, pointed out that AIDS screening was a "waste of time."[126] Dr Blatherwick observed: "One might want to point out that Canadians have a higher rate of HIV infection than most other countries. If we're going to screen anyone, maybe we should screen Canadians going out of the country. We are more part of the spread of AIDS than some poor unfortunate person coming in."[127] While Canada was unsure about following the example of its southern neighbour, US President Bill Clinton's attempt to lift the ban preventing foreigners with AIDS from immigrating to the country was defeated in Senate. On 18 February 1993, the US Senate voted to maintain the ban and change its nature from policy to federal law.

Meanwhile in Canada, on 26 April 1994, the *Toronto Star* reported that Immigration Minister Sergio Marchi had confirmed Ottawa's intention to test all potential immigrants, thus ignoring the opinions of experts who had concluded the measure was unnecessary. Talking to reporters, Marchi said, "Yes, we are looking at whether HIV testing should be automatic."[128] The minister explained that the issue was under review "because of the potential strains on the health-care system that could be caused by admitting immigrants with HIV,"[129] though he could not say whether the ruling would apply to refugees. The government was under pressure; Reform MP Art Hanger asked the minister to "scrap a federally funded international AIDS conference, scheduled for 1996 in Vancouver, because it will bring about 500 visitors infected with HIV into Canada."[130] Mr Hanger also sponsored a private member's motion "to make HIV testing mandatory for all immigrants and refugee claimants and automatically deny entry to those who test positive."[131] The government came under attack from the Reform Party after granting refugee status to a Polish man infected with AIDS. The man, who died in January 1995, had arrived in Canada four years earlier, "claiming that he faced persecution because he was both a homosexual and HIV-positive."[132] The Immigration and Refugee Board accepted the claim, but Reform MP Philip Mayfield asked the minister "to reverse the board's decision and deport the man."[133] In the House of Commons, Marchi defended Canada's decision arguing that the "case does not set a precedent."[134] Professor James Hathaway, expert in refugee law at Osgoode Hall, agreed that "the case is unlikely to provoke a flood of either homosexual or HIV-positive refugee claimants because claimants must demonstrate they face serious harm in their own country and that the state is unwilling or unable to stop the abuse."[135] And yet, the image of hordes of destitute refugees banging at Canada's doors continued to haunt most ordinary Canadians, thus offering further validation to those politicians whose goal was to keep out undesirables.

In the meantime, other cases unrelated to HIV/AIDS were keeping immigration authorities occupied. On 2 September 1994, the *Globe and Mail* published the story of Mohamed Mussa, a forty-nine-year-old refugee from Somalia who was denied permanent resident status because he used a wheelchair after having lost a leg in a car accident in his home country. Although he had full-time employment and required "almost no extra health or social services because of his paralysis, the federal government perceives Mr. Mussa as a potential drain on resources."[136] The article noted that under new regulations expected to take effect in October, it was going to be "even easier for the government to keep out – and deport – people like Mr. Mussa because of their medical condition."[137] The new legislation was based on a revised handbook for medical officers listing illnesses and disabilities together with the costs to health and social services. Immigration officers were instructed to reject any applicant whose projected cost over five years was greater than what an average Canadian would incur in the same period of time.[138] The new rule was introduced to eliminate any discretion on the part of overseas medical officers and to restrict the entrance of people with either illnesses or disabilities, resulting in "savings to health and social services."[139] Immigration lawyers and disability rights groups criticized the legislation because of its failure to take into account immigrants' cultural backgrounds, thus assuming that every disabled immigrant was going to overuse social services or health facilities – even though "many cultures consider such a move inappropriate or even taboo."[140] These comments echo those made by experts two years earlier in the parliamentary committee created to analyse Bill C-86, a Measure to Amend the Immigration Act.[141] In that venue, professionals presented evidence indicating that on average immigrants tended to place lower demands on health and social services than Canadian citizens. Lawyers also feared that increasing rejections due to medical admissibility was going to result in more cases going "to costly years of appeals."[142] Finally, as noted by Vancouver immigration lawyer Dennis McRea, the model omitted consideration of the long-term financial benefits of the applicant's family since "if they were contributing a few thousand dollars in taxes to the system, and their disabled member was only using a few hundred dollars more than the average Canadian in health and social services, it [would be] a good economic position for the government."[143] Mr McRea's comments perpetuate the erroneous belief that, while the family as a whole will contribute to Canadian society, the disabled member will inevitably remain a burden. Scholars have repeatedly pointed out that these are "unfounded and discriminatory assumptions"[144] based on outdated economic and biomedical models of disablement. When provided with proper accommodation, most disabled individuals are able to work. More important, it is reductive and demeaning to assess human beings exclusively in terms of their economic contributions.

Rejection of family members with a disability was especially contentious in the case of children. The courts were sceptical of decisions barring children on medical grounds. On 6 February 1996, the *Toronto Star* reported that the Federal Court, in the person of Justice James Jerome, overturned the immigration department's decisions denying permanent resident status to two families whose children had been diagnosed with mental disabilities. According to the judge, it was wrong to use "mental retardation" as a stereotype and "to judge the children as if they were adults applying for landed immigrant status."[145] Lawyer Cecil Rotenberg commented, "the key to winning in court is to prove that the family will not let the child become a drain on the public purse."[146] This approach was rejected by federal authorities; Dr Neil Heywood, director of immigration health policy, explained, "once families are given landed status, they have the freedom to ... use whatever services are available to Canadian citizens."[147] I will discuss this aspect of the law in the next chapter.

Organizations representing disabled persons considered the Immigration Act in need of revision since, under the guise of economic concerns, it discriminated on the basis of disability and therefore appeared to be in violation of the Charter. As argued by Diane Richler and reported by the *Globe and Mail*, "Most branches of law, guided by the Charter, now regard disabilities as a human-rights issue and not a medical concern."[148] A report published in 1999 and prepared by an umbrella group of thirty-four non-governmental organizations labelled Canada "hypocritical for outlawing discrimination against people with disabilities while explicitly permitting such discrimination in immigration and refugee cases."[149] The issue was particularly troublesome in the case of children since Canada's rejection of immigrant children with disabilities resulted in the systematic violation of the UN Convention on the Rights of the Child, "which was adopted by the United Nations in 1989" and ratified by all but two countries, "the United States and Somalia."[150] Canada's violation of international laws is an interesting subject that would require a book of its own. For my purposes though, it is important to keep in mind that the issue brought to the forefront the complexity of reconciling the country's commitment to the international community with questions of self-interest.

The new century witnessed a continuation of the debate surrounding immigration of persons with HIV/AIDS. In September of 2000, Citizenship and Immigration Minister Elinor Caplan revealed that acting on a recommendation from Health Canada, the department was going to ban immigrants with HIV from Canada. The minister mentioned that an exception would be made for refugees and immigrants with close family members already in the country. The rationale behind the ruling was that persons with HIV "could put a strain on Canada's health care system and ... endanger other residents of Canada."[151] The announcement was met with criticism. Raif Jurgens of the Canadian AIDS-HIV Legal Network[152] noted that according to the UN, "there are no public health grounds for HIV testing of immigrants."[153] He added, "HIV is very different from tuberculosis. It is not easily transmitted. They are singling out HIV because it's politically more palatable, because [people who have it] are suffering from discrimination already."[154] Indeed, HIV-positive persons "might not develop AIDS symptoms for years, and ... they could contribute to society in ways that would outweigh health-care costs."[155] Dr Philip Berger, AIDS expert at St Michael's Hospital in Toronto, explained that the HIV virus was not casually transmitted and therefore "people can protect themselves."[156] The move smacked of racism because it targeted immigrants from specific countries: "Since HIV prevalence is higher in African countries, this would be a good way of keeping out immigrants from those areas. This policy will discriminate against people in the poorest countries in the world – mostly non-white countries."[157] Several readers disagreed. In a letter published in the *Star* on 26 September 2000, Ms Yetman of Wolfville, NS, expressed support for the new policy: "As a nation we need to be primarily concerned with the safety and health of our current visitors, landed immigrants and Canadian citizens ... Do we really want to put more money into an already deteriorating health-care system in order to treat people who have not (yet) contributed to the tax system and who, by virtue of carrying the virus, may not be well enough to ever contribute?"[158] The *Globe and Mail* came out strongly in favour of the policy. Criticizing the Human Rights Commission for condemning the ban against HIV immigrants, the paper reminded its readers:

> Health Canada ... estimates that among those 200,000 new arrivals about 200 are HIV-positive, and that together they will infect a further 37 individuals at some future date. One issue is the expense of treating HIV patients with the drug cocktail that slows the progress of AIDS. The annual cost for a patient can be as high as $10,000. More important, AIDS still has no cure, and unless one is found, those 37 freshly infected people will face the prospect of a ghastly, premature death ... Health Canada's initial proposal for HIV testing would provide an extra element of protection.[159]

Fortress Canada was the solution to this foreign menace.

Paradoxically, Health Canada was not so sure about its own recommendations. In what the *Star* termed "a dramatic reversal,"[160] Health Minister Allan Rock sent a letter to his counterpart in the immigration department informing that "his department has decided to 'refine' its advice and no longer advocates an outright ban on HIV carriers."[161]

Mr Rock explained, "An inflexible policy of excluding all immigrants who test positive would not reflect the nature of the risk nor our capacity to minimize it."[162] Because of the change in Health Canada advice, Citizenship and Immigration Minister Caplan announced that the ban was to be reversed. Yet, the department retained the power to reject an immigrant with AIDS if the person posed an excessive demand on Canada's health-care system.[163] On 13 June, a letter appeared in the *Toronto Star* condemning the reverse of the ban as "ridiculous."[164] According to the writer, Mr Paul Taylor of Scarborough: "We already let in people who bring no skills, aren't healthy, not to mention those who slip into Canada who are convicted, or wanted, criminals from other countries. If you are a health risk, you are a health risk: You should not be allowed to enter Canada no matter what the circumstances. The decision to let individuals in who are HIV positive to eliminate the stigma that is too often attached to those living with HIV is pathetic."[165] Evidently, the belief that Canada was and could remain a safe haven immune from certain diseases still found fertile ground among ordinary readers.

The year 2002 opened with a case that summarizes the entire debate surrounding medical admissibility. On 8 January, the *Toronto Star* published an article written by staff reporter Maureen Murray chronicling Angela Chesters, a native of Germany married to a Canadian citizen who was denied permanent resident status because she was diagnosed with multiple sclerosis (MS). Angela and Robin Chesters married in 1991 after meeting in Frankfurt where Robin was working at the time. They decided to move to Canada in 1994, but immigration authorities classified Angela as inadmissible. Mrs Chesters decided to challenge the legislation. Ena Chadha, senior counsellor at ARCH,[166] explains that the case was "the first of its kind to make its way to the Federal Court."[167] As usual, Ottawa tried to settle the case out of court by granting Mrs Chesters permanent resident status on an individual basis, but she refused the offer arguing that it did not go far enough to change the law and showed signs of charity. Although Ottawa had recently passed amendments to the legislation "exempting spouses and dependent children from having to submit to the test of whether they would be a burden to the health and social services system,"[168] Angela Chesters continued her fight in the belief that "a person should not be banned from immigrating to Canada on the basis of their disability."[169] Debra McAllister, counsel representing the federal government, dismissed the claim that persons with disabilities were discriminatorily excluded. Ms McAllister argued that "people are not excluded simply because they have MS or are in a wheelchair"[170] and asserted that each case was assessed on an individual basis. She added, "For the protection of Canadian society, immigration officials need the ability to deem some people inadmissible if they are likely to pose an excessive burden."[171] Mrs Chesters disagreed, insisting that the immigration department had looked exclusively to her disability without taking into account possible contributions she could have made to Canadian society. As a matter of fact, it refused to consider that "she holds master's degrees in history and science, is fluent in French and has broad work experience."[172]

Instead of quietly accepting the government's offer of a minister's permit, Mrs Chesters "launched an unprecedented court bid to have the Canada Immigration Act declared unconstitutional because it allows discrimination on the basis of disability."[173] During cross-examination, Ms McAllister confronted Angela Chesters. When Ms McAllister, referring to the offer for a ministerial permit, asked: "Everything Canada could have done it did, correct? It did everything possible,"[174] Mrs Chester replied, "Incorrect."[175] And when Ms McAllister showed her a neurologist's report indicating she was likely to require increasing nursing care in the future, Mrs Chesters responded, "With all due respect, you may require increasing nursing care."[176] The government's claim that applicants were not assessed solely on the basis of their disability was discredited by the testimony of Dr Ted Axler, retired medical officer from Canada Immigration who reiterated that assessments considered "how much a person will cost the heath-care system and ... how employable they are on 'generic' and 'stereotypical' factors."[177] Dr Axler told the court that employability is "generalized; it is not based on an individual's work record but on the likelihood of people with a certain illness being employable."[178] He added that excessive demand consisted of "estimates of the costs of disorders ... based on expectations for all patients with a disorder, not individuals."[179] Dr Axler put forward the example of internationally renowned physicist Stephen Hawking who "would be denied permanent residency status in Canada because he suffers from a debilitating neurological disease."[180] Indeed, "if we were lucky enough to have an application for landing from Professor Stephen Hawking, we would have to say that this international treasure would be inadmissible."[181] Undoubtedly, Dr Axler chose an example that had a good chance of resonating well with his audience. Martha Rose remarks that while the average able-bodied person is never questioned about her/his worth, the idea that disabled people "must make extraordinary contributions to society in order to be worthy" is widely circulated in North American public discourse.[182] Why should a person with a disability be expected to go over and above the "norm"? And why are extraordinary contributions only required from persons with disabilities? This double-standard approach is questionable, particularly when it comes from a country that has built its reputation around the principles of freedoms, rights, and equality.

Outside of the court, Angela Chesters's case animated a lively debate between supporters and opponents of the existing legislation, and it received unprecedented attention in the press. Readers were touched by her story and sent letters of support. One of these letters, published in the *Globe* on 9 January, maintained: "Canada should be honoured that someone so accomplished, with such a resolve to be a contributing member of society, would want to live and work here ... Whatever 'excessive demands' she may require in the future, she will more than adequately give in return with her talents, not the least of which is her unfailing spirit and dignity as a person, despite a debilitating illness."[183] The *Toronto Star* showed its support in an editorial on 10 January: "Whether the court decides in Chesters' favour or not, the way she was treated is an embarrassment. It contravenes Canadian values and basic human decency. Chesters' manifest ability to contribute to the country was judged less important, by immigration officials, than the possibility she might someday need nursing care."[184] On 1 July, after the federal court ruled against Angela Chesters, the paper observed:

> Canada's immigration department is the face we show the world. It can be a harsh one. Angela Chesters ... was ruled inadmissible because she might make excessive demands on Canada's health-care system ... But Angela Chesters refused to leave without a fight. She took the immigration department to court on the grounds that it had violated her constitutional right to freedom from discrimination based on physical disability. This past week, the Federal Court of Canada ruled against her ... For the immigration department, this is a sweet legal victory. For Canadians who believe in fairness, tolerance and basic decency, it is an embarrassment.[185]

If nothing else, the case proved that the interests of Canadians did not necessarily coincide with the interests of their government.

Many things could be said to conclude this chapter. Among them, the fact that all the excerpts from newspaper articles in this chapter as well as the previous one indicate that a large portion of the Canadian public often ignore, or rather prefer to ignore, what goes on around immigration. Better to assign that responsibility to bureaucrats and politicians, only to blame them when a sensational story reaches the headlines. After all, it is easier to deal with problems on an emotional level rather than address the systemic unfairness of a legislation that has been kept essentially unaltered for more than a century. Showing pity is easier than initiating a serious discussion around rights and our complicity in silencing these rights. In *Looking White People in the Eye*, Sherene Razack makes this point when she challenges her readers to move beyond pity to respect, thus beginning the process of questioning our involvement in the oppression of others.[186] When we pity someone, we tend to focus on becoming saviours who can bring salvation to the unfortunate of this world, but we fail to account for our role in creating the situation that resulted in a victim in need of rescue. If we are really committed to an antioppressive and anticolonialist future, we must begin the conversation by looking at how we are participating in subjugation.

Another important and related finding in this chapter is that articles, letters, and editorials analysed reveal unevenness in the attitudes manifested by Canadians towards those deemed inadmissible to the country, particularly in relation to race. The evidence provided in this chapter validates the argument I made in the conclusion of the previous chapter and suggests that there was undoubtedly a divergent approach among readers in their dealings with, on the one hand, cases like the one of Angela Chesters, an educated, white, middle-class woman with MS, and, on the other, cases concerning persons with HIV or AIDS, especially when they were coming from non-Western countries. While most readers were unanimous in their support for the admittance of applicants who, except for their disability, were considered "normal" and could easily integrate into Canadian society, the attitude remained negative when dealing with persons whose illness suggested an anomalous lifestyle, one condemned as unhealthy and deviant, particularly in the case of racialized individuals. As noted in the previous chapter, I agree with Parin Dossa that attitudes towards race, gender, and disability, far from being separate domains, tend to operate in conjunction and strengthen each other.[187] In this sense, readers and writers seem to show a greater acceptance and empathy towards certain applicants over others, thus using race as a discriminatory tool for assessing membership within the national body.

It is also important to bear in mind, as evidenced by the comments reported in the press, that authorities as well as the public continued to assess immigrant applicants on the basis of their productivity: whether healthy or unhealthy, no person was welcomed to Canada if deemed unproductive. Refugees represented the only partial exception,[188] though the articles analysed in this chapter have highlighted that, once accepted, refugees were often unfairly blamed for abusing the system. Canada receives refugees because of its international responsibilities, but this does not prevent criticism and grumbling on the inside. Under neo-liberal restructuring, Canada has increasingly reframed the concept of citizen and transformed it into what Dossa refers to as "citizen worker"[189] – someone who receives validation as a member of the community only insofar as they are able to function within the market economy. Those who do not meet the criteria are systematically marginalized and devalued.

In this chapter, I have examined articles, readers' letters, and editorials published in the *Toronto Star* and the *Globe and Mail* in the period from the passage of the Charter to 2002. Although dealing with the same topic at the core of the previous chapter, I decided to dedicate a separate chapter to an examination of the press in the period following the passage of the Charter because it has deeply influenced public discourse in Canadian society. Since the Charter's institution, both newspapers adopted a different tone in their reporting on the cases of medically inadmissible immigrants. Whereas the period preceding the mid-1980s saw both papers maintain overall support for the medical admissibility clause, in spite of sporadic outbursts of sympathy for individual situations, the content of contributions published from that date onward indicate a more nuanced approach. From the late 1980s, the press called for a revision of the section of the Immigration Act dealing with medically inadmissible persons, particularly in situations where the application of the law resulted in the splitting of families. This change validates the argument that the passage of the Charter has to a certain degree impacted Canadian society.

In conclusion, and before I move to investigating the discourse formulated by the courts, I would like to end the chapter by returning to Angela Chesters's reply to Debra McAllister in the Federal Court in 2002. When asked, "Everything Canada could have done it did, correct? It did everything possible,"[190] Mrs Chesters answered, "Incorrect."[191] Indeed it was and is incorrect. Despite opposite claims, Canada refuses to accept disabled or ill immigrant applicants. It refuses to consider the opinion of experts in the scientific community who have repeatedly indicated that the legislation is based on faulty assumptions. More important, Canada refuses to consider immigrants as human beings rather than mere producers. The concept of economic burden that resonates so often in the public discourse presented and legitimized by the press does not merely reveal the non-place of immigrants with disabilities within Canadian society, it also exposes the way immigration in general has been among the main tools used to shape the national body according to principles of usefulness and productivity.

5 Medical Admissibility in the Federal and Supreme Courts of Canada

This chapter examines several court cases related to the medical admissibility provision in the Canadian Immigration Act of 1976. The cases reveal the state policy on medical admissibility and provide evidence that the shifts registered in politicians' and newspapers' discourses are a smokescreen: different language is used, but ultimately the decision of excluding undesirables stands. In previous chapters, I argued that the passage of the Charter, in particular its section 15 (the equality clause), represented a milestone in the recognition of rights for persons with disabilities in Canada. Sarah Armstrong observes that the Charter "became the first constitution to guarantee a right to equality for persons with disabilities."[1] Since the passage of the Charter, advocacy groups and individuals have been able to use the document in a court of law to advance their claims. In the 1989 decision in *Andrew v. Law Society of British Columbia*, the Supreme Court of Canada ruled that the Charter's protection extends to persons who are neither Canadian citizens nor permanent residents. Justice McIntyre "maintained that non-citizens are a minority group analogous to those enumerated under the grounds of s. 15 and that they come within the protection of s. 15."[2] The decision opened a new avenue for immigrant applicants to fight those provisions of the Immigration Act barring them from Canada. In this chapter, I argue that the Charter has the potential to substantially impact the medical admissibility provision. Whereas claims of discrimination by private individuals or groups are usually handled under statutory human rights laws, discrimination by government (including laws and policies) is under the purview of the Charter.[3] Given the significance of the document, it is useful to reflect on how it has been brought into play in the last decades by advocacy organizations and those affected by the medical admissibility provision.

In order to investigate the role that the Charter played in questioning medical admissibility, this chapter examines seventeen cases brought to the Federal Court and the Supreme Court of Canada by plaintiffs contesting the immigration department's decision to refuse them or their family members entrance into the country. It is difficult to categorically establish whether these cases constitute the majority of the legal actions presented in courts since not all cases get reported and those that do are often selected because electronic search engines make them easily accessible. Furthermore, whereas litigation challenging the law on Charter grounds is usually reported because, involving the Charter, it is considered noteworthy, other lawsuits would simply go unchallenged and end up with the person being denied entrance. I believe that the cases analysed here are illustrative of the courts' standing. Legal decisions follow the rule of *stare decisis* (let the previous decision stand), and this is evidenced by the fact that judges constantly reference previous court decisions in their orders. If any additional relevant case exists on the topic, it is unlikely that the courts would have omitted referencing it. Throughout my investigation, I have not discovered any case that challenges the general trend I delineate through the seventeen court cases discussed below.

The earliest lawsuit among those examined in this chapter was heard in May 1988, while the latest was heard in February 2005. Although this book covers the years from 1902 to 2002, a few court cases that were heard after 2002 have been included as they appealed previous court decisions or their outcome was significant for the purpose of the discussion. Two of the cases reached the Supreme Court; all others were assessed in the Federal Court of Canada.[4] At the outset, I expected that most of the cases would directly engage with section 15. After all, the equality clause is the first to come to mind when dealing with issues of discrimination by the government against individuals belonging to equity-seeking groups. I was therefore bewildered to find out that only one of the seventeen cases made explicit reference to the Charter, while the remaining cases contested the decision without questioning the existing law, merely arguing that the provision did not apply to the specific situation under review. Despite my initial puzzlement, an in-depth analysis reveals that there is a subtle connection between the way the cases unfolded and the Charter. It appears likely that faced by the argument that the medical admissibility clause does not refer to persons with disabilities, but focuses instead on the concept of excessive demand, the plaintiffs were forced to shift their arguments away from disability and prove instead that they were not likely to cause excessive demand on Canadian health and social services. In so doing, while contesting in a tangential way the discriminatory nature of the legislation, the core of their arguments eluded the issue. The only exception was the hearing of Angela Chesters. In that instance, Mrs Chesters maintained that the provision was unconstitutional since it offended section 15 of the

Charter by discriminating against persons with disabilities. Although unsuccessful, Angela Chesters's hearing is extremely important for understanding the potential impact the Charter might have on the way the Canadian state relates to medically inadmissible applicants. For this reason, the analysis of the legal action initiated by Mrs Chesters in 1997 is near the end of the chapter, followed by a broader discussion around the results of the Charter's application in the courts. However, it is worth clarifying that the focus of this chapter is on the process of exclusion of immigrants with disabilities rather than on a single litigation, notwithstanding its unquestionable uniqueness. In this context, the Chesters case becomes a means to an end rather than an end in itself.

The first court case considered is *Pamar v. Canada (Minister of Employment and Immigration)*. It was heard in the Federal Court of Appeal on 3 May 1988. The appellant, Swaranjit Kaur Pamar, contested the refusal of the sponsored application presented by his mother and her accompanying dependants. Mr Pamar's mother had been declared inadmissible under subparagraph 19(1)(a)(ii) of the Immigration Act of 1976 on the grounds that she suffered from ischemic heart disease, exercise-induced myocardial ischemia, hypertension, and non-insulin-dependent diabetes mellitus, conditions that were assessed by Canada as likely to cause excessive demand on health and social services. According to Mr Pamar, the decision was based on the unreasonable opinions of the medical officers. The appellant argued, "the diagnosis upon which the expectation that the person's admission to Canada will create excessive demand on Canadian health services does not flow reasonably from the medical evidence."[5] On 16 May 1988, Justice Heald dismissed the appeal since there was enough medical evidence in support of "the conclusion reached by the medical officers that Mrs. Pamar's admission to Canada would create excessive demands."[6] Justice Heald noted that, as required by law, the opinion of the medical officer had been confirmed by a second medical officer. Four additional medical reports had been taken into consideration, three of them supporting the findings of the medical officers. The court therefore sided with the Immigration Appeal Board and resolved that there was reasonable and grounded evidence for rejection.

A different outcome was reached a few years later in the Federal Court of Appeal in *Deol v. Canada (Minister of Employment and Immigration)*. The appeal was directed against the decision of the Immigration and Refugee Board and sought to dismiss a previous appeal from the immigration officer's refusal of a sponsored application. In that circumstance, the appellant's mother and her two dependant daughters were refused admission on the grounds that one of the daughters was diagnosed as mentally retarded and could reasonably be expected to cause excessive demand on health and social services. The appellant contested the board's decision on the basis of both validity and equity. In the judgment delivered on 27 November 1992, Justice MacGuigan allowed the appeal and ordered the matter returned to a differently constituted panel for rehearing and redetermination. While recognizing that the board cannot question a medical diagnosis, MacGuigan declared, "the Board did not inquire into the reasonableness of the medical officers' conclusion, but rather assumed ... that the conclusion was reasonable."[7] MacGuigan added: "The mere invocation of mental retardation leads to no particular conclusion. Mental retardation is a condition covering a wide range of possibilities from total inability to function independently to near normality. The concept cannot be used as stereotype, because it is far from a univocal notion. It is not the fact alone of mental retardation that is relevant, but the degree, and the probable consequences of that degree of retardation for excessive demands on government services."[8] As expected, the judge's argument did not question the law but merely the errors committed in its application.

In contesting the board's decision, the appellant also argued that the board was wrong in establishing that the onus of proof was on the applicant. Given that mental retardation is a form of mental disability and therefore is an enumerated ground of discrimination under section 15(1) of the Charter, Deol maintained that any justification had to be made under section 1 of the Charter, which reads, "The *Canadian Charter of Rights and Freedoms* guarantees the rights and freedoms set out in it subject only to such reasonable limits prescribed by law as can be demonstrably justified in a free and democratic society."[9] Deol claimed that the onus was on the government. Justice MacGuigan rejected the argument since it contradicted section 8(1) of the Immigration Act, which explicitly placed the burden of proof on the person seeking entrance into Canada. Nevertheless, the court allowed the appeal on the basis that the board failed to take into account the nature and degree of the mental retardation as well as humanitarian and compassionate considerations, such as "close bonds of affection that may arise in such a family."[10] In so doing, the board had ignored the intention of the Immigration Act to facilitate family reunification. *Deol v. Canada* is significant insofar as the appellant made a direct albeit minor reference to section 15. The court acknowledged that the issue had never been argued in front of the court. However, Justice MacGuigan decided that the contention was inconsequential, thus implying that section 8(1) of the Immigration Act had precedence over the Charter. While the Charter is an integral part of the Canadian constitution and is therefore, as stated in section 52(1) of the Constitution Act, the most important law in the country, the courts have often refused to acknowledge the incongruity between the Charter and several of Canada's laws. Particularly with respect to immigration, the courts' unwillingness to examine whether immigration policy conforms to principles of universal justice allows the state to operate under what Voyvodic defines as a "wasteland."[11] Similarly, Giorgio Agamben refers to "state of exception," a blind juridico-institutional spot where the state can ignore its own laws in the name of necessity.[12] According to Agamben, this dynamic occurs because, by having the power to decide what to include as well as what to exclude in the law, the sovereign (in this case, the state) is by definition above the law itself and can therefore decide if and when exceptions to the law can be made.[13] As evidenced by the examples in this book, the state continues to reject immigrant applicants with disabilities although the exclusion appears to violate its legislated commitment to non-discrimination on grounds of disability. The courts are complicit in this process of discrimination as failing to acknowledge such violation only helps to legitimize it.

In 1995, another lawsuit reached the Federal Court Trial Division. Marcel Gingiovenanu, a Romanian citizen who had applied for permanent residence in Canada, submitted an application for judicial review of a visa officer's negative decision. The initial application had been refused because Mr Gingiovenanu's son had cerebral palsy and was therefore assessed as inadmissible under section 19(1)(a)(ii) of the Immigration Act. The applicant claimed that the visa officer's decision was unreasonable since his son was not expected to require surgery or institutionalization. Justice Cullen dismissed the application on the grounds that the medical officer's conclusion was "not unreasonable because it flows from the preponderance of the evidence."[14] Notwithstanding the dismissal, the court reaffirmed that visa officers have a responsibility to consider whether medical assessments are reasonable. They have to take into account whether a medical condition can arguably be assumed to create excessive demand. As evident in the cases that follow, the courts have traditionally been inclined to limit the authority of medical officers in the matter of rejecting immigrants. The courts' stance rests on the conviction that denying admission to Canada for medical reasons is an eminently political decision and, as such, cannot be left to medical officers. Final decisions remain the direct responsibility of the Department of Citizenship and Immigration.

A few months earlier, Justice Cullen had taken a similar approach in *Ismaili v. Canada (Minister of Citizenship and Immigration)*. Mr Ismaili was refused an application for permanent residence under the assisted relative class on the basis that his son, diagnosed with severe developmental delay and a bilateral hearing deficit with a history of seizures and microcephaly, was likely to pose excessive demand on Canadian health and social services. The court allowed the application, arguing that the visa officer had failed to consider whether the medical officer's decision ignored evidence. Referring to *Deol v. Canada*, Justice Cullen reiterated that "a medical condition, alone, is not necessarily evidence of a reasonable expectation of excessive demands on health and social service" and that "the visa officer – wholly apart from the decision of the medical officers – is obliged to consider whether the applicant's medical condition would place excessive demands on health or social services."[15] Based on the evidence in the record, the court was not convinced that the visa officer had considered all the factors. The court found that the applicant's son was going to require inexpensive medication and special schooling not in high demand in the region of Canada selected by Mr Ismaili for settlement. Justice Cullen concluded that these requirements did "not amount to excessive use"[16] and allowed the application.

The question of how visa officers should assess medical opinions was at the centre of *Poste v. Canada (Minister of Citizenship and Immigration)*. In 1997 John Russell Poste submitted an application for judicial review of a decision denying permanent resident status to the Federal Court Trial Division. A former Canadian citizen who lost his citizenship after moving to Australia in 1973, Mr Poste had subsequently applied to return to Canada as a permanent resident together with his Australian wife and children. The application was denied because the eldest son had a mental disability. The applicant contested the decision as unreasonable in light of the fact that his wife was a nurse and could take care of the child. Mr Poste pointed out that his son, Matthew, was entitled to an Australian pension while living within or outside Australia, and hence was not going to make excessive demands on health and social services in Canada.

Looking at the evidence before the court, Justice Cullen agreed that the medical officers' opinion was unreasonable since they had reached a conclusion without considering the specific circumstances of the case. In delivering the court's order, the judge stated: "There is an absence of evidence to support the conclusion of excessive demand. There is no evidence to show that the medical officers put their minds to the question of excessive demand as it relates specifically to Matthew. To the contrary, the evidence seems to show that the medical officers only considered the demands placed on social services by the mentally disabled in general. The medical officers have a duty to assess the circumstances of each individual that comes before them in their uniqueness."[17] The court concluded that the medical officers had not met the requirement for individual assessment and denounced the visa officer's decision as unreasonable. The visa officer had indeed accepted the medical officers' opinion without question and rejected Poste's application. Justice Cullen found this approach problematic: "The visa officer must not simply accept a medical officer's determination of medical inadmissibility as the basis for rejecting an applicant's AFL [application for permanent status]. To do so effectively gives medical officers 'carte blanche' authority to decide who can immigrate to Canada. The final decision must rest with the visa officer, who has a duty to assess all the circumstances of the case."[18] The visa officer adduced that one of the reasons for rejecting the application was the fact that Matthew was unlikely to ever become autonomous and financially independent. Justice Cullen dismissed this reasoning as inconsequential and noted, "Matthew was applying under the category of a dependent ... There is no requirement under the Immigration Act for a dependent to establish self-sufficiency."[19] The court allowed the application for judicial review since

> it would be a "win-win-win" situation if the applicant and his family were allowed to immigrate to Canada. The first "win" would be that Canada would be gaining two new professionals, resourceful people: the applicant and his wife. The second "win" would be that, although the applicant's son, Matthew, has a mild mental disability, he would be accompanying his family as, by all accounts, a well-adjusted individual considering his circumstances ... The third "win" would be in furtherance of Canada's policy of family reunification. The applicant ... has an elderly mother in Canada who needs his help.[20]

In the Poste case, the judge's focus was entirely on the benefits that would accrue to Canada by accepting the family. This reasoning is in line with the previously discussed mainstream understanding of immigration as a tool meant to assist Canada.

The requirement of considering non-medical factors when assessing a person's likelihood to cause excessive demand was reiterated in *Lau v. Canada (Minister of Citizenship and Immigration)*. In 1998, Hing To Lau appealed to the court for judicial review of a visa officer's refusal of his application for permanent residence under the investor category on the grounds that one of his daughters, Kwan, was assessed as medically inadmissible. Kwan had a moderate mental disability and was deemed likely to cause excessive demand on social services. Mr Lau alleged that the decision was unreasonable since his daughter would reside with him and therefore had family support. Both the medical officer and the visa officer provided affidavits to the court and were cross-examined on their testimonies. The medical officer explained that he had reached a negative conclusion since he could not be sure that family support was going to continue into the future.[21] A similar rationale was provided by the visa officer who stated that "we have no guarantees. They may have plans for now, but we don't know what may happen."[22] Justice Pinard allowed the application as "the lack of due consideration of 'family support' constitutes a blatant failure to consider all of the evidence with respect to the personal circumstances in the applicant's file."[23]

While stating that family support was among the factors to consider when assessing if a person was likely to cause excessive demand, Justice Pinard did not clarify whether family support extended to financial assistance, thus leaving the matter open to interpretation by individual judges. The question resurfaced in *Poon v. Canada (Minister of Citizenship and Immigration)*. In 2000, Ching Ho Poon applied for judicial review of a visa officer's decision that refused him and his family immigrant visas after his son, Tat Chi, was diagnosed as moderately retarded and likely to cause excessive demand on health and social services. Poon testified that he was willing to underwrite the cost of all services required by Tat Chi. Justice Pelletier dismissed the argument on the basis that Poon's wealth and willingness to pay for services were irrelevant. Referring to a 1995 court order in *Choi v. Canada (Minister of Citizenship and Immigration)*, Pelletier clarified that "the applicant's argument that the family's ability to pay has not been considered, cannot succeed ... Access to health and social services in Canada is a matter of right for citizens and permanent residents. Once Tat Chi became a permanent resident, he would be entitled to claim access to such publicly-funded services as he required and any agreement to the contrary would be unenforceable against him."[24] Although rejecting the contention that parents' financial assistance should be taken into consideration, the court allowed the application and ordered the matter remitted to a different visa officer for redetermination. The court found the medical report insufficient because it only took into account the cost of services while failing to consider their supply. Justice Pelletier explained that "some consideration must be given to the supply of services in order to conclude that demand would be excessive. I am unable to find any consideration of supply in the record."[25] By relying on invalid medical opinion, the visa officer had made an error of law.

Meanwhile, the fairness of the opinions formulated by medical officers continued to be debated in court. In 2001, Jeannine Elise Redding, a US citizen, applied to the Federal Court Trial Division for judicial review of a decision refusing her application for landed status. Ms Redding held two postgraduate degrees from Canadian universities and was working in Canada with an employment visa. Her application was rejected because she had juvenile diabetes. The applicant contested the decision and cited medical reports indicating that her attention to diabetic care contributed to stabilizing potential complications, giving her an excellent prognosis. Justice Lemieux concluded that the medical officer had ignored Ms Redding's individual circumstances and reached the decision by exclusively looking at the general diabetes population. Referring to the court decision in Lau, Justice Lemieux reiterated that assessments in evaluating medical admissibility had to be made on an individual basis. In the case of Ms Redding, the medical officer failed to perform such an assessment: "The applicant's material shows that her condition is stable and her prognosis is excellent. Dr. Waddell [the medical officer] simply does not deal with her situation."[26] The court allowed the application and returned the material to a different visa officer and different medical officers for redetermination.

The question of whether a person's ability to pay for social services ought to be considered when assessing medical admissibility came up again in *Wong v. Canada (Minister of Citizenship and Immigration)*. In 2002, Ching Shin Henry Wong applied for judicial review of a visa officer's decision denying his application for permanent residence under the self-employed category due to the fact that his daughter had been diagnosed with Down's syndrome. Wong had been refused two applications in 1994 and 1996, but he had successfully applied for judicial review in the second instance. However, the medical officers had again reached the conclusion that the girl was medically inadmissible since she was likely to place excessive demand on social services. Wong contested the decision arguing that family support had not been taken into account and that the family was willing to pay for all social services needed. While the court found that the medical officer had indeed considered family support, the application was allowed because the officer had failed to note that the social services required in the case under review were not covered by the government in the province selected by the applicant as place of residence. Accordingly, the girl's admission was not going to cause excessive demand. Recalling previous court decisions, Justice McKeown asserted:

> The jurisprudence is split on the question of whether the wealth of the applicant should be taken into account in assessing excessive demands on social services. While in Ching Ho Poon v MCI [2000] F.C.J. No. 1993 (T.D.) Pelletier J. found that wealth was not relevant, in my view the better approach was taken by Reed J. in the earlier Wong decision when she found that it would be incongruous to admit somebody as a permanent resident because he has significant financial resources but then refuse to take into account these same resources when assessing the admissibility of the dependant. This approach would not be applicable in the case of medical services but it is applicable with respect to social services.[27]

The judge further explained: "In Canada one is not permitted to obtain medical services on a private basis. However, there is no such restriction in the social services and as was shown by Ontario's Development Services Act, persons who can afford to pay for social services must pay for them."[28] The court's conclusion in *Wong v. Canada* became a launch pad for similar lawsuits. It remains debatable whose interests are foregrounded in these cases. Due to the high costs of litigation, it is safe to assume that only a restricted number of affluent applicants possess the means to follow through on a lengthy court battle. The majority of those affected by the legislation are likely unable to cover the costs. Moreover, few immigrant applicants have enough money to pay for social services once in the country. Discrimination on the basis of wealth is not covered in the Charter.

According to the Medical Services Act passed in 1968, Canadian citizens and permanent residents have the right to publicly funded coverage of visits to hospitals and doctors as well as to diagnostic services (referred to as medical services). Some social services might also be covered depending on the province of residence, particularly when such services are deemed to be an important component in the management of certain health conditions. By allowing into Canada only those immigrants with disabilities who can cover their expenses, the notion of citizenship is reframed in accordance with market imperatives.[29] I am critical of the shift that is taking place around citizenship. Countries that have embraced a neo-liberal political discourse and governance framework have indeed gradually emphasized their reliance on the new concept of "consumer-citizen"[30] that constructs subjects as deracialized, degendered and declassed but not demonetized. I particularly fear the way this discourse is gradually eroding any understanding of citizenship as an inclusive status while replacing it with the idea of membership based exclusively on economic participation and competition. In the field of immigration, we still have to conduct a comprehensive investigation on the consequences of considering wealth as among the determining criteria for admission within the country.

A few months later, a major lawsuit heavily relying on *Wong v. Canada* was brought forward by David Hilewitz. The applicant, a citizen of South Africa, asked for judicial review of the visa officer's rejection of his application for permanent residency in the investor category. The application was refused because one of his children, Gavin, was diagnosed with developmental delays and assessed as inadmissible under section 19(1)(a)(ii) of the Immigration Act. Although the IRPA had been passed in 2001, section 190 of the new act established that the case had to be decided on the basis of the repealed Immigration Act since the initial application for permanent residency had been refused when it was still in effect. Mr Hilewitz argued that the visa officer's conclusion that Gavin was going to cause excessive demand on social services was incorrect since the family was wealthy enough to pay for any required service and intended to do so. Looking at the evidence presented and at previous court decisions, in particular *Wong v. Canada*, Justice Gibson concluded, "I am satisfied that the officer erred in a reviewable manner in her failure to consider all of the material available to her."[31] The appeal was allowed. The decision of the visa officer was set aside and the application for permanent residence referred to a different officer for redetermination.

Following the court decision, the minister appealed under the argument that the trial division judge had erred in stating that the parents' wealth and willingness to pay for social services were relevant when deciding on admissibility. The Court of Appeal acknowledged that "in his excessive demands opinion, the medical officer did not take into account Mr. Hilewitz' financial means"[32] and that "there is considerable Trial Division case law supporting the proposition that ability and willingness to pay are relevant considerations."[33] Nevertheless, the court concluded that the minister was correct in arguing that financial considerations were irrelevant when assessing medical admissibility. The minister's appeal was allowed, the verdict of the applications judge was set aside, and the initial decision of the visa office restored. Justice Evans noted that, according to section 19(1)(a)(ii)

> the excessive demands on health and social services to be considered are those that are "a result of the nature, severity of probable duration of" the disability. This would seem to limit the factors about the individual on which the medical officer is to base an excessive demands opinion to the diagnoses and prognoses of the medical condition, and the health and social services that are thereby likely to be required, but not such non-medical considerations as an ability and willingness to pay for needed services. In addition, the fact that Parliament entrusted responsibility for forming an excessive demands opinion to medical officers may also suggest that they were not intended to have to take into account non-medical factors ... that are not within their expertise.[34]

Furthermore, even if non-medical factors were to be considered, "once admitted to Canada visa applicants ... may relocate to a place where publicly funded social services are available without cost recovery ... or where the services required are not available privately. Finally, there is also a speculative aspect about predicting the life choices that a person with a disability may make ... Gavin may decide in the next few years to try to make as independent a life for himself ... Financial misfortune or some other unforeseen change of circumstance may also prevent the family from providing the material support for which they had planned."[35] Justice Evans recognized that his interpretation of the law as mandating the medical officer to exclusively consider medical factors seemed in contradiction with the objective of the Immigration Act to facilitate the admission of those meeting the other entry qualifications in a particular category, especially when they were people of wealth who were likely to contribute to the Canadian economy. In fact, "It could be said that Canada's ability to use immigration policy to attract capital and entrepreneurial talents may be unduly hampered by the exclusion of a person who is expected to make significant contribution to the Canadian economy, if medical officers are not required to conduct a full assessment."[36] The court observed the problem could be addressed by issuance of a ministerial permit[37] that functioned as a probationary admission, thus enabling the department to reassess the situation after three years. Furthermore, it was Parliament's responsibility to find a balance between the benefits of having someone likely to make a significant economic contribution and the risk that the admission would result in excessive demand on social services. Determining how risk-averse Canadian immigration policy should be was Parliament's choice to make.

Justice Evans's argument seemed to be in response to the criticism of scholars who disapproved of judicial activism and accused the court of bypassing the democratic process, thus reversing policies made by elected governments.[38] Such criticism has been particularly intense in the field of Charter jurisprudence. Allan C. Hutchinson, for example, characterizes judges and lawyers as "a professional elite that can have no claim to constitutional priority over democratic deliberation."[39] According to Hutchinson, constitutional supremacy has resulted in judicial supremacy, thus bringing Charter issues out of the "political forum of democratic debate and into the legal arena of judicial pronouncement."[40] The most detrimental outcome of this process is a decrease in active participation among the population, with citizens transformed into passive subjects.[41] Although in a more nuanced form, Manfredi appears to share this view when he suggests that "we might be better off rediscovering the value of the public realm of politics"[42] rather than expecting the judiciary to solve all of society's problems around issues of liberty, equality, and social justice. While containing elements of truths, especially with respect to the contentions that any legal interpretation is political and that so far courts' decisions have not been particularly progressive,[43] the argument against juridical intervention looks weak. I find Hutchinson's call for popular action as the only avenue for achieving political and social change disingenuous.[44] His celebration of popular uprisings such as the French revolution in 1789 and Timisoara in 1989 ignores that social change can also occur through other means. More important, Hutchinson's call for "dialogic democracy,"[45] an alternative space where tensions can be recomposed through negotiation between the parts, fails to give due consideration to the fact that the players involved have different degrees of power, making the playing field decidedly uneven. Whereas I do agree with Dossa that Canada's liberal democracy tends to quash actions insofar as people are persuaded that once their rights are enshrined in the Charter, "they are implemented as a matter of course,"[46] this should not result in neglecting the reality that the Charter has improved the life of individuals belonging to equity-seeking groups.

A similar approach to the one taken in *Hilewitz v. Canada* had been adopted the year before by the Federal Court Trial Division in *De Jong v. Canada (Minister of Citizenship and Immigration)*. De Jong, a citizen of the Netherlands, applied for judicial review of the visa officer's rejection of his application for permanent residence in the self-employed category. The application was rejected because De Jong's dependant daughter was diagnosed as having a developmental delay and was expected to cause excessive demand on social services. The applicant argued that the visa officer failed to consider his financial resources, which would provide for his daughter's special needs. He maintained that in light of the distinction between health and social services whereby in Canada only health services are publicly funded across the country, wealth was a factor to consider when assessing excessive demand on social services. Referring to the previous decision rendered by the Federal Court of Appeal in *Deol v. Canada*, yet ignoring the conclusions reached by Justice McKeown in *Wong v. Canada*, Justice Pinard concluded that the distinction between health and social services was irrelevant when deciding on medical admissibility.[47] Referring to the judgment passed by the Federal Court of Appeal, Pinard noticed that the court had made reference to "any services required,"[48] which disregarded the distinction between health and social services. Accordingly, he dismissed the application. De Jong appealed the decision to the Federal Court of Appeal. The appeal was heard in 2003, immediately after *Hilewitz v. Canada (Minister of Citizenship and Immigration)*. In that circumstance, the court held that medical officers were not legally obliged to consider non-medical factors – such as the applicant's wealth – when assessing the likelihood of excessive demand on health and social services. Abiding by those principles, the court dismissed De Jong's appeal.

In 2005, Hilewitz and De Jong appealed the decisions reached by the Federal Court of Appeal to the Supreme Court of Canada. This time, both appeals were successful, and the applications were referred to the minister of citizenship and immigration for reconsideration and redetermination by a different visa officer. In rendering the court's judgment, Justice Abella held that both visa applicants qualified for admission to Canada because of their assets. For that reason, it appeared "incongruous to interpret the *Immigration Act* in such a way that the very assets that qualify these individuals for admission to Canada can simultaneously be ignored in determining the admissibility of their disabled children."[49] According to the wording of the medical admissibility provision:

> Section 19(1)(*a*)(ii) calls for an assessment of whether an applicant's health would cause, or might reasonably be expected to cause excessive demands on Canada's social services. The term "excessive demands" ... shows that medical officers must assess likely demands on social services, not mere eligibility for them. Since, without consideration of an applicant's ability and intention to pay for social services, it is impossible to determine realistically what "demands" will be made, medical officers must necessarily take into account both medical and non-medical factors. This requires individualized assessments. If medical officers consider the need for potential services based only on the classification of the impairment rather than on its particular manifestation, the assessment becomes generic rather than individual ... Given their financial resources, H [Hilewitz] and J [De Jong] would likely be required to contribute substantially, if not entirely, to any costs for social services provided by the province of Ontario, where they wish to settle.[50]

The court dismissed the minister's argument that the applicants' abilities to pay for those services could not be taken as a guarantee for the future. Indeed, "the fears articulated in the rejections of the applications, such as possible bankruptcy, mobility, school closure or parental death, represent contingencies that could be raised in relation to any applicant"[51] and were therefore invalid.

The court noted that the initial legislation on the matter of immigration to Canada had allowed for the entrance of people who could demonstrate enough wealth to ensure they would not become public charges. An absolute ban was only introduced in 1927. Responding to "concerns that such policies were overly restrictive,"[52] the Immigration Act enacted in 1976 introduced the excessive demand standard as a replacement to the wholesale rejection of prohibited classes. Justice Abella noted: "The issue is not whether Canada can design its immigration policy in a way that reduces its exposure to undue burdens caused by potential immigrants. Clearly it can. But here the legislation is being interpreted in a way that impedes entry for *all* persons who are intellectually disabled, regardless of family support or assistance, and regardless of whether they pose any reasonable likelihood of excessively burdening Canada's social services."[53] The refusal to consider an individual's ability and willingness to absorb these costs was deemed contrary to the original intent of the act. With respect to the applications submitted by Hilewitz and De Jong, "the medical officers were obliged to consider all relevant factors, both medical and non-medical, such as the availability of the services and the anticipated need for them. In both cases, the visa officers erred by confirming the medical officers' refusal to account for the potential impact of the families' willingness to assist."[54] Abella did not question Canada's reasoning regarding selecting immigrants on the basis of economics, and he did not even contemplate the possibility that this determination might conflict with the constitution of the country. I maintain that, by failing to consider potential conflicts between certain Canadian laws and Canada's constitution, the courts are abdicating their responsibility.

Two of the judges in the panel, LeBel and Deschamps, dissented with the decision and argued that wealth was not among the factors that should be taken into account when assessing medical admissibility. Both judges sustained that based on the legislation,

> the determination of excessive demands is made by reference to the nature, severity and probable duration of the medical condition itself. Rather than exclude persons on the basis of the condition alone, Parliament intended the medical officer to look at how the condition affects the individual. This does not, however, mean looking at criteria that have nothing to do with the medical condition ... the fact that Parliament expressly considered whether family support was relevant to excessive demands assessments and chose not to include it in the *Immigration Act* and the regulations strongly suggests that Parliament did not intend wealth to be a relevant factor.[55]

Besides, obliging medical officers to consider non-medical factors was problematic since "the more a medical officer's analysis is tied to highly subjective non-medical factors, the more likely it is that the medical officer will be drawn into assessments outside his or her area of expertise."[56] While acknowledging the apparent incongruity of accepting visa applicants because of their assets yet refusing to consider those same assets when determining excessive demand, LeBel and Deschamps concluded that "this is what Parliament has done. It has chosen to use criteria for the decision on medical inadmissibility that are distinct from those used for the selection as business or economic applicants. Business or economic applicants are evaluated on the basis of their potential contribution to Canada; however, in order to avoid undermining their potential contribution, these applicants must not fall into an inadmissible class of persons. The applicants can still be admitted on the basis of their wealth, but this is left to the discretion of the Minister who can issue a permit despite the medical inadmissibility."[57] However, the minister's prerogative to make exceptions was not shared by medical and visa officers. Over time, there had been no indication that wealth was among the factors officers were obliged to consider: "When wealth was to be considered by the decision maker, the statutes said so clearly. Section 19(1)(a)(ii) makes no such reference. In its historical context, this silence is meaningful."[58]

According to Justice Deschamps, the omission of wealth as one of the factors to consider was indicative of the will of Parliament:

> Neither s. 38(1)(*c*) of the subsequent statute, the *IRPA* [Immigration and Refugee Protection Act], nor its accompanying regulations (see s. 34 of the *Immigration and Refugee Protection Regulations*, SOR/2002-227) make any reference to family support or income; instead, s. 34 directs officers drawing conclusions about excessive demands to consider only reports made by a health practitioner or medical laboratory and any condition identified by the medical examination. Likewise, the handbook given to medical officers to assist them in making their assessments specifically directs them to ignore "civil factors, such as the economic circumstances of the applicant" and to focus "solely on the medical considerations specified in the Act and Regulations ... If Parliament had wanted to direct medical officers to consider family support or wealth, it had ample opportunity to do so when revising the rules. It is not for the courts to make such revisions in a case where there is no constitutional challenge.[59]

The fact that the 1976 Immigration Act and its successor omitted any reference to wealth in the question of medical admissibility was not accidental. The legislation was "not drafted in a vacuum; it is impossible to ignore the context of widespread health and social safety nets which existed when the provision came into force."[60] It was fair to assume that Parliament had acted with the specific intention of protecting the financial assets of the state from depletion.

The conclusions reached by the Supreme Court in *Hilewitz v. Canada* and *De Jong v. Canada* set a dangerous precedent by legally establishing that financial ability to pay for social services is important when assessing medical admissibility. It bears the implication that visa applicants should be assessed depending on their wealth. In so doing, it validates the distinction among people on the basis of financial means and economic potential while ignoring other non-economic contributions. With respect to the declared objectives of the IRPA, it aims "to support the development of a strong and prosperous Canadian economy, in which the benefits of immigration are shared across all regions of Canada"[61] while brushing aside the social and cultural benefits that immigration brings to the country. As I have suggested in previous chapters, the decision confirms that the main purpose of Canadian immigration policy is to support the country's economy through the admittance of a healthy and productive workforce. Under the assumption that only productive subjects are useful and that disabled individuals are inherently unproductive, it also sends a message to Canadians that this standard is the ideal they should embody. Failure to attain that ideal will not put them outside the legal framework of citizenship; nonetheless, it will make them less worthy. I find it problematic when society reduces its members to mere producers and consumers, valuing them exclusively based on the money they possess. I find it even more troubling when a court of justice validates such an argument. Ireh Iyioha interprets the reasoning of the court as based on "procedural fairness," namely the "need for an individualized cost assessment based on the individual."[62] Yet, in adopting a neo-liberal perspective on the meaning of citizenship and participation in the community, the court outright ignored "the imperatives of the legal right to equality and non-discrimination inherent in both international human rights law and the Canadian Charter."[63] The narrative embraced by Canada's courts contributes to legitimizing the replacement of the universal (at least in its intentions) notion of citizen with the new one of "consumer-citizen"[64] whose value is determined by the person's buying power.

The argument that wealth should be the main factor in deciding admission to the country was further validated when almost two years later Canada Immigration and Citizenship (CIC) finally responded to the Hilewitz and De Jong rulings and released Operational Bulletin 037 outlining new procedures complying with that ruling. As immigration lawyer Guidy Mamann points out: "In its usual 'never-give-an-inch' fashion, the department outlined a more individualized procedure, but limited it 'only to applications made under the business class'" ... all applicants who are in similar circumstances and who are applying in other immigration classes are out of luck."[65] Consequently, the Department of Citizenship and Immigration decided that immigrants with disabilities were to be assessed exclusively on their financial means. Only applicants considered wealthy – especially those applying under the business class category, a category that is no longer in existence since it was cancelled by the Conservative government of Stephen Harper – got a chance to prove their ability to pay for services. Others were denied the same opportunity. This development was quite incongruous with respect to the equality clause's statement that "every individual is equal before and under the law."[66] Despite its attractiveness, the concept of equality remains challenging in its definition, and interpretations of the term equal vary in time and according to context. As an aside, it is worthy to mention that a search on the web has shown no reference to Operational Bulletin 037 other than an announcement on the Citizenship and Immigration Canada's website[67] and Mamann's article that was first published in *Metro*, a free daily publication available in Toronto, Vancouver, Montreal, Ottawa, Calgary, Edmonton, and Halifax;[68] it was then reproduced on the website of Mamann's immigration law practice.[69] It would be worthwhile to examine why major newspapers, particularly the two examined here, ignored the document and its implication for the immigration policy of the country. Their choice is indicative of how press coverage is selective in its nature. We could speculate that the *Star* and the *Globe* deemed the policy change of little relevance because it did not affect any particular individual and was therefore unlikely to generate interest among the Canadian public. There was no damsel in distress to save from the dragon – it was merely another bureaucratic ruling.

As illustrated by the court cases discussed above, almost all of the immigration lawsuits brought to court avoided direct reference to the Charter. Instead of questioning the law as potentially unconstitutional because it was in violation of the equality clause, the reviewed litigations simply cast doubts on the correctness of applying the medical admissibility provision to the specific circumstances of the appellants. It is curious that at a moment when "the Charter has invigorated the struggles of groups that represent people with disabilities,"[70] references to the Charter were absent in the overwhelming majority of legal cases dealing with individuals who had been deemed medically inadmissible. It is possible this occurred because legal counsels were of the opinion that questioning the constitutionality of the Immigration Act would be unsuccessful in court. Aside from the traditional conservatism of the courts,[71] there were two fundamental obstacles to overcome: first, the government's claim that the issue at stake was not disability but excessive demand created by disability, and second, the fact that section 1 of the Charter establishes that rights and freedoms guaranteed by the Charter are not unconditional but subject to reasonable limits (and what is or is not reasonable is a matter of interpretation) prescribed by the law. Furthermore, Evelyn Kallen remarks, "individuals or organizations bringing complaints under the Charter must pay for the costs involved."[72] Recognizing that costs can be prohibitive, especially for persons belonging to disadvantaged minorities, the federal government had initially set aside some funding for "selected 'Charter Challenge' court cases at the federal level. However, this program was discontinued in 1992."[73] Eliminated by the Conservative government under Brian Mulroney, the program was revived by the Liberals when they came to power in 1993. After being criticized for years by social conservatives and critics of judicial activism for providing groups with the money to go to court and thus circumvent the will of Parliament, the program was killed once again in the fall of 2007 by Stephen Harper's Conservatives. As a result, many persons have been prevented from pursuing the option of using the Charter in the protection of their rights. The high costs involved in bringing Charter claims forward have made the document inaccessible to those needing it most. If not for the subsidies provided by organizations advocating on behalf of equity-seeking groups, as well as the benevolence of some lawyers to bankroll cases, the Charter would remain a luxury many cannot afford.

Despite these difficulties, in 2001, as discussed earlier, Angela Chesters, a German citizen married to a Canadian, challenged as unconstitutional the visa officer's decision to refuse her permanent resident status under the family class category on the basis that she had MS and was likely to cause excessive demand on health and social services. Her case represents the first serious attempt to question the medical admissibility provision as unlawful because in violation of section 15. As shown in the previous chapter, it was also noteworthy because of the publicity it received in the press and the debate it generated across Canadian society. The fact that the effort to prove the unconstitutionality of the provision did not succeed is indicative of Canadian law's understanding of immigration and disability.

In December 1991, Angela Chesters married Robin Chesters, a native of South Africa who had subsequently acquired Canadian citizenship. In September of that year, Mrs Chesters had been diagnosed with MS and by the end of 1993 was using a wheelchair. At that time, the couple lived in England. In 1994, they decided to move to Canada and Mrs Chesters submitted an application for permanent residence in the family class. On 14 November 1994, Robin Chesters received a telephone call from Mr Ernest Alston, a visa officer at the Canadian High Commission in London where Angela had submitted her application, informing him that his wife was medically inadmissible. Mr Alston advised Robin Chesters to investigate the possibility of obtaining a ministerial permit. By that time, the couple was living in Canada since Mrs Chesters had received a visitor's visa in the summer 1994. In April 1995, Angela Chesters was issued a ministerial permit valid for three years. The following year, she moved back to Germany and her husband joined her in February

1999. After receiving a negative response from immigration authorities, Mrs Chesters did not pursue an application for judicial review but challenged the constitutionality of section 19(1)(a)(ii) by way of action. Her objection was formalized through the issuance of a statement of claim, a court document used in Canadian law to start a court case, on 2 April 1997 (an amended statement was filed on 29 July).[74]

On 30 August 2000, the minister of citizenship and immigration made an offer in writing to settle the proceeding, which included landing for the plaintiff. On 27 September 2000, Mrs Chesters rejected the offer arguing that it was made to silence her. On the basis that the offer had been declined, the court dismissed the minister's claim that the issue before the court was moot, as there was no live controversy between the parties to the litigation. The rationale behind the request for dismissal was that the court had competence to resolve legal disputes within an adversarial system and this could not be done without the existence of a true adversarial context. Nevertheless, the court has authority to hear a case when the latter involves collateral consequences for the litigants. The court can also refuse dismissal when there is an intervener with a stake in the outcome of the litigation. Justice Dawson concluded there were valid reasons for the court to hear the case. The plaintiff's request was for a declaration that section 19(1)(a)(ii) was inconsistent with section 7 and section 15 of the Charter: "While at the end of the day the plaintiff certainly may hope to be landed (and I note parenthetically to be landed by right and not by the defendant's largesse as reflected in a settlement offer), it remains that the plaintiff's action does not directly put landing in issue and is not limited to a claim for landing."[75] In December 2001, the defence asked for summary judgment and repeated its request to have the issue declared as moot in reason of the fact that a new Refugee and Protection Act had been given royal assent on 1 November 2001, though it had not yet come into force. The minister pointed out that Mrs Chesters "would no longer be medically inadmissible under subsection 38(2) of the new legislation."[76] As previously noted, the new act establishes that the medical admissibility provision does neither apply to refugees nor to spouses, common law partners, or children of a Canadian citizen or permanent resident. Justice Lufty dismissed the motion on the grounds that "the plaintiff's action is grounded on the current legislation, not on the new *Immigration and Refugee Protection Act* which has yet to come into force. In his letter to the Court dated December 4, 2001, counsel for the plaintiff stated that the new legislation has no impact on this litigation."[77] Incidentally, should the new legislation apply, the motion submitted by the defence would still be problematic. Leaving aside the issue of refugees, which would require a separate investigation, granting immigration status to spouses and children of Canadians was a decision made to guarantee that the law conformed to one of its objectives – family reunification.[78] However, the problem with the change in the legislation is that spouses are only assessed as appendages of a Canadian citizen / permanent resident, and only because the latter is considered worthy of the right to choose a companion without the state interfering. Within this framework, the applicant is not considered as a person with intrinsic worth but as "the spouse of," a quite demeaning position to be in. Of course, defenders of the legislation might argue that the situation is not a novelty and that it is in fact common to all those applying as members of the family class. However, there is ground to contest this interpretation. Whereas the family category is by definition assessing individuals as part of the family unit, I think we need to question the fact that while all adults have a choice whether to apply as independent or sponsored, that choice eludes persons with a disability or a disease. They are discriminated against insofar as they are presented with no real alternatives in the application process.

In rejecting the motion to have the case dismissed for mootness, the court considered the fact that, by order dated 20 February 2001, Prothonotary Lafreniere[79] had granted intervener status to the Council of Canadians with Disabilities. On 27 February 2001, the CCD made a written submission to the court arguing that it was important to adjudicate on the constitutionality of section 19(1)(a)(ii) since the provision contributed "to the pervasive negative stereotype of persons with disabilities, including those who are permanent residents of Canada, as a drain on society."[80] For years the Department of Citizenship and Immigration avoided having the constitutionality of the act adjudicated in court by issuing ministerial permits to people who had initiated a Charter challenge. The consequence was that "those who are desperate for permanent status in Canada usually do not have the resilience to see a case through its conclusion when tempted by the blandishment of a Minister's Permit."[81] The CCD insisted that "the constitutionality of section 19(1)(a) of the Act, is regarded by the CCD as an issue of national importance. This is not only because of its impact on persons with disabilities from abroad, who would be welcomed as new Canadians were it not for their disability; but of equal importance is its discriminatory impact upon Canadians with disabilities who are stigmatized by this legislation's failure to recognize that disabled people make contributions to Canadian society."[82] Whereas I do not necessarily endorse the CCD position considering I have demonstrated that the situation of Canadians with disabilities is inherently different from that of disabled immigrant applicants, I agree that, at least at the level of perceptions, it is credible that some Canadians with disabilities might feel demeaned by legislation banning disabled foreigners.

Angela Chesters's hearing reached the trial stage on 27 June 2002. In her submissions to the court, Mrs Chesters argued that she was discriminated on the basis of her medical condition, "which was improperly categorized by Canadian immigration officials as a disability."[83] She challenged the constitutionality of section 19(1)(a)(ii) on the grounds that it contravened section 7, which guarantees life, liberty, and security of the person, and section15 of the Charter. Section 7 was violated because she was placed in a situation of uncertainty resulting in high levels of mental and psychological stress. Section 15 was breached as she had been singled out as belonging to a class of persons who were explicitly protected by that section. She maintained that even "if the provision is not discriminatory on its face, then it still breaches the guarantee of equality because it has an adverse discriminatory effect."[84] The CCD submitted that the medical admissibility provision was "further flawed by failing to address the potential contribution to be made to Canada by persons suffering from disabilities."[85] As an aside, I found it quite disconcerting that the CCD, whose mandate is to advocate on behalf of Canadians with disabilities, chose the word "suffering" in referring to persons with disabilities. This choice of vocabulary is a huge setback. Tanya Titchkosky writes that by adopting the rhetoric of sufferance to describe the connection between people and disability, we objectify the nature of disability as "a condition to be suffered – clearly and obviously a big problem."[86] The result is an understanding of disability as necessarily bad. While working to improve the lives of persons with disabilities, the CCD is simultaneously validating those same perceptions of disability that are at the root of the devaluation of disabled persons.

In presenting its case, CIC argued that section 19(1)(a)(ii) did not offend section 7 and section 15 of the Charter and was, in any event, justifiable under section 1. Canada had both the right and obligation to set standards for entry into the country, and subparagraph 19(1)(a)(ii) was meant to accomplish that objective by protecting Canadian health and social services against excessive demand. CIC denied Mrs Chesters's claim that the department's actions had resulted in a breach of section 7 of the Charter since levels of stress are to be assessed objectively and that section was never intended to "protect an individual from the ordinary anxiety that would be felt by a person of reasonable sensibility, as the result of government action."[87] As for section 15, CIC relied on a previous decision reached in 1999 by the Supreme Court in *Law v. Canada (Minister of Citizenship and Immigration)*, where the court identified three steps to be followed in assessing whether section 15 was breached. The first step was to look at an appropriate comparator group receiving a different treatment from the plaintiff. Angela Chesters argued that a distinction was made between able-bodied and disabled spouses of Canadian citizens who applied for permanent residence, yet CIC countered that the distinction was between persons who were medically admissible and those who were not. The second step considered whether the distinction was based upon the enumerated grounds of disability. CIC stated that the distinction had no relation to disability and was instead based on the concept of excessive demand. Finally, the third step looked at discrimination based on stereotypical reasoning. CIC argued that Mrs Chesters had been "assessed on a personalized basis"[88] and there had been no stereotypical reasoning behind the visa officer's rejection.

In giving the reasons for the court order, Justice Heneghan stated that "it is well-settled law that not every differential treatment gives rise to discrimination."[89] In assessing whether discrimination had taken place, he relied on the three-step process outlined in *Law v. Canada (Minister of Citizenship and Immigration)*. On the question of a proper comparator group, Justice Heneghan concluded that Mrs Chesters had sought admission to Canada as member of the family class; hence, the family class and not able-bodied spouses of Canadian citizens was the appropriate comparator group. Accepting CIC's argument, the court interpreted section 19(1)(a)(ii) as focusing on excessive demand rather than disability: "Subparagraph 19(1)(a)(ii) is not directed to any of the specified grounds in subsection 15(1) of the Charter. It is directed to excessive demand."[90] The court explained that the article had the purpose of protecting Canada against excessive demand and "that purpose is apparent even if the words 'disease,' 'disorder' or 'disability' are deleted."[91] Justice Heneghan's reasoning for choosing the family class as comparator group appears questionable. Dale Gibson explains that, under Charter jurisprudence, the "ambit of the comparison group is ... no longer capable of being defined by the law itself"[92] (in this case, the Immigration Act); rather, it depends on "who is complaining and what they are complaining about."[93] Obviously, Angela Chesters was not complaining about the law itself but the way it was applied to a subgroup of the applicant population, namely persons with disabilities. In *Andrew v. Law Society of British Columbia*, Justice McIntyre remarks that to ignore that a similar treatment can result in the discrimination of certain individuals[94] is deceitful at best. By looking at the cases examined in this chapter, it is apparent that the application of the law resulted in singling out a specific group of individuals who all were diagnosed either as disabled or diseased. To claim this was a coincidence is simply absurd.

Observing that entry into Canada is a privilege rather than a right, Justice Heneghan concluded:

> The section in question focuses on excessive demand, not on disease, disorder or disability. Contrary to the stance taken by the plaintiff, this case is not about disability but the medical assessment of potential immigrants to Canada within the context of Canadian immigration law. By its nature, legislation governing immigration must be selective ... The process of assessing medical examinations for the purpose of determining excessive demands upon existing Canadian health services is an aspect of the screening process ... this is not within the enumerated grounds of subsection 15(1) nor is it analogous to it.[95]

With respect to Mrs Chesters's claim that the actions of the state had put her in a situation of uncertainty resulting in high levels of psychological stress, thus contravening the guarantee of life, liberty, and security of the person under section 7 of the Charter, the court dismissed the argument. Noting that "the constitutional guarantee of security of the person does not protect against 'ordinary stresses and anxieties that a person of reasonable sensibility would suffer as a result of government action,'"[96] the court concluded, "the plaintiff's right to security was not infringed in consequence of the application of the inadmissibility section. She was not the victim of state action. As a potential immigrant, she was subject to meeting the requirements of the *Immigration Act* and Regulations."[97] Accordingly, "subparagraph 19(1)(a)(ii) offends neither subsection 15(1) of the Charter nor section 7."[98]

In giving the reasons for its decision, the court did not address any of the submissions presented by the intervener, the CCD. It failed to address the intervener's criticism that the medical admissibility provision did not balance the potential contributions to Canadian society of persons with disabilities or illnesses against the expected costs to health and social services. On that issue, Justice Heneghan chose to remain silent. Sarah Armstrong observes that several scholars have repeatedly concluded that the judicial system is inherently conservative and the courts are more likely to uphold the status quo than validate a progressive interpretation of the Charter.[99] This conservatism of the judicial system occurs in particular when decisions are related to immigration.[100] Catherine Dauvergne claims that "migration decision-making" is characterized by both "executive discretion and judicial deference," which is demonstrated by "the scant impact which the Canadian *Charter of Rights and Freedoms* has had on immigration and refugee law."[101] The court decision in the case of Angela Chesters confirms Dauvergne's interpretation.

Nevertheless, in a number of cases, the courts have played a significant role "in safeguarding minority rights overlooked in the ordinary processes of majoritarian democracy."[102] While the assumption that "once the rights of people are enshrined in Canada's Charter of Rights and Freedoms, they are implemented as a matter of course"[103] is patently untenable, the Charter represents a valuable tool in the advancement and protection of minority rights within the judicial system. With reference to individuals with mental or physical disabilities, their inclusion in section 15 of the Charter has represented a victory in their struggle for recognition.

This inclusion did not come as a gift but was instead fought for by the disability community. When section 15 was first introduced in Parliament in October 1980, it made no reference to persons with disabilities. The exclusion was strongly opposed by the disability community on three different fronts: in the media and in public debates over patriation; in discussions of the Special Committee of the House of Commons on the Disabled and the Handicapped, also known as the Parliamentary Handicap Committee; and in presentations to the Special Joint Committee of the Senate and House of Commons on the Constitution of Canada (the Hays-Joyal Committee).[104] The disability community advocated for the constitutional recognition of the right to equality for persons with disabilities, while the federal government resisted the idea of having mental and physical disability added to section 15. Minister of Justice Jean Chrétien argued that there were three main objections to including the disabled among the protected groups: first, the courts were already in a position to intervene if a person with disability was discriminated; second, there was no clear definition yet in Canadian society of who were those covered by the term disability; finally, there was no need to protect disability rights under the constitution as the provinces already had effective human rights codes.[105]

Another objection, never clearly stated but always hinted at, was that it would be too expensive to fund new services that might be required by the Charter. However, "at no point did the government present statistical or factual support for the claim that including disability in section 15 would cost too much money."[106] Before the Hays-Joyal Committee, several organizations such as the Coalition of Provincial Organizations of the Handicapped (COPOH),[107] the Canadian Association for the Mentally Retarded (CAMR), and the Canadian National Institute for the Blind (CNIB) contested the government's position by arguing that "disability was the only ground of discrimination to which this cost-benefit analysis was being applied."[108] Excessive costs have also been used to justify the medical admissibility provision in the Immigration Act of 1976 and in the IRPA. Iyioha refers to several research studies indicating that despite the persistence of the common view that immigrants are depleting Canadian health care resources, they are actually underutilizing them.[109] In fact, government claims around the likely expenses involved in accepting a certain category of immigrants are largely unsubstantiated. Of course, I am in no way trying to validate the narrative behind such claims. Even if immigrants took advantage of Canadian health care resources at the same rate the rest of the population does, I would not consider the finding a valid excuse to restrict their access. Human beings are more than the money they make or the services they use. Our society has a serious problem if it only assesses members based on financial considerations.

In the fight to have disability included in the Charter, it played to the advantage of disability rights' advocates that the debate took place during the UN International Year of Disabled Persons. Canada had been one of the co-movers of the UN resolution that designated 1981 as the International Year of Disabled Persons.[110] Given these premises, it became difficult to justify the refusal to entrench the rights of persons with disabilities into the new constitution while at the same time supporting the principles of full participation and equality entrenched in the UN resolution.[111] In the end, the government capitulated and "Canada added both mental and physical disability to the list of specifically protected groups in sections 15(1) and 15(2) of the Charter."[112] Providing persons with disabilities with constitutional recognition of their right to equality magnified their impact on society. They found themselves in possession of a new legal tool in their quest for equality. Armstrong points out that "since the enactment of the Charter, groups representing people with disabilities have repeatedly intervened in equality rights cases to persuade the courts to adopt this vision of substantive equality."[113] Although the judicial system has not always embraced a progressive view of the Charter (and here the term progressive refers to the active protection of minority rights that are overlooked in the ordinary process of majoritarian democracy), it has been empowering to have a voice in the court.[114] Armstrong adds that "the potential for interest advocacy under the Charter has given disability groups significant new opportunities to influence the development of the law and public opinion on fundamental disability issues."[115] This latter point is quite significant because, as we have witnessed over and over, little happens in terms of legislation unless sustained societal demands for action are directed at institutions of power.

This chapter has illustrated how the medical admissibility provision was contested in the courts. It has further revealed the state policy on medically inadmissible immigrants, thus proving that – despite the changes in language registered in politicians' and newspapers' public discourses – there has been no fundamental ideological shift in the way immigrants with diseases or disabilities have been assessed in the last century and a half. The discussion has shown that regardless of the language used to justify the exclusion, whether employability or excessive demand, the decision to reject individuals who are disabled or diseased has continued to stand. The chapter has also focused on the impact of the Charter in the interpretation of the provision. In reason of the conservatism of the courts and CIC's argument that rejection on the basis of medical admissibility is not about disability per se but excessive demands on state resources, the overwhelming majority of the litigations examined in this chapter failed to use the Charter to question the constitutionality of the law. They merely contested the appropriateness of applying the law in the particular instances considered. From the analysis of the Chesters hearing, it also transpires that the courts have so far been reluctant to address questions of unconstitutionality, which confirms the arguments of various scholars who maintain the judicial structure tends to keep the existing system intact. It would be worthwhile to conduct a thorough examination of the reasons behind such attitudes, beginning with considering to what extent this judicial conservatism is a reaction to accusations repeatedly addressed to the courts that complainants are attempting to bypass the democratic process and reverse Parliament's decisions.

The Charter's application in the courts is no panacea. The judiciary tends to have a conservative approach to the equality clause and appears inclined to uphold the status quo. Nevertheless, the Charter represents a tool in the hand of minority groups. With respect to medical admissibility, the applicability of the Charter to every person who is on Canadian soil, citizens and non-citizens alike, has a potential that still needs to be fully explored. It brings back to the stage the issue of human rights guaranteed to every individual irrespective of nationality. In 1951, in *The Origins of Totalitarianism*, Hannah Arendt claimed, "general human rights, as distinguished from the rights of citizens ... proved to be unenforceable."[116] Arendt's argument has recently been echoed by Giorgio Agamben who maintains that the notion of human rights "is inconceivable in the law of the nation-state,"[117] and that there are no rights outside those bestowed upon citizens. Catherine Frazee, drawing on Ignatieff's work, remarks that the notion of universal human rights keeps clashing with the instinctual understanding that we should only care about people like "us."[118] The Charter represents an attempt to invalidate such arguments. Whereas still tied to the notion of the nation state,[119] the fact that the

Charter applies to every individual physically in Canada without consideration for their nationality renders it an opportunity of unprecedented proportions.

The refusal of the court in 2002 to engage with the criticism brought forward by the CCD is visible proof that many of the implications of the Charter have been disregarded so far. However, these same issues will likely resurface, and the court will have to confront them again. Although I am not a legal scholar, I do believe that the admissibility provision could potentially be found unconstitutional because in violation of the Charter. Indeed, it is revealing that in the only case dealing with the issue of unconstitutionality, the court refused to address the inconsistency of the provision vis-à-vis the Charter while limiting itself to argue that the provision was not targeting disability but excessive demand. As shown, there is considerable ground to contest this claim and demonstrate that the provision is not only about excessive demand but also, and to a significant degree, about disability. Once the intrinsic connection between the provision and disability is established, the court will have to revisit its conclusion. This time the outcome could be – and I hope it will be – quite different.

Conclusion

In this book, I have argued that Canadian immigration policy regarding persons with a disease, disorder, or disability has historically been discriminatory. I validated this argument through the analysis of the public discourse entertained by federal politicians, two major newspapers (the *Toronto Star* and the *Globe and Mail*), and the court system in the years from 1902 to 2002. This conclusion summarizes my main findings, takes stock of major changes over time, and briefly hints at some of the challenges ahead. It also discusses how the discrimination embodied in the medical admissibility provision impacts on the notion of citizenship in Canadian society, transforming it into a tool of exclusion rather than inclusion.

One of the goals of this book was to attempt to fill the gap in the existing scholarship – considering that issues of immigration, disability, and citizenship have often been discussed separately. The previous chapters demonstrated that the way selected sectors of Canadian society – namely federal politicians, various contributors to two widely read newspapers, and the legal system – discussed disability was reflected in their approaches to immigration and their understandings of citizenship. The historical devaluation of persons with disabilities in Canadian society has resulted in the search for ideal citizens who are healthy and productive, to the exclusion of everyone outside those parameters. While several studies have separately examined the Canadian immigration process, the situation of persons with disabilities, and the construction of citizenship in the Canadian context, few investigations have explicitly linked these elements. The disconnect among these fields has resulted in the failure to understand what Sherene Razack calls "interlocking oppressions."[1] People's racial, gendered, and other identities are not separate compartments but various facets that coexist, interact, and interlock with one another to position groups and individuals within the national context or outside of it.

Failure to establish a connection among immigration, citizenship, and disability within the Canadian context has also hampered a full understanding of the processes regulating the formation, development, and maintenance of a Canadian national identity. As Catherine Dauvergne indicates, especially in countries of immigration such as Canada, the United States, Australia, and New Zealand, "the mythology of the nation is bound up with immigration ... when these nations select immigrants ... they convey the nation's values."[2] A full understanding of Canadian society cannot avoid considering the vital role migration laws have played in the construction of the nation as these laws represent the main instrument for the selection of community members. Whether such laws include or exclude persons with disabilities, and whether they position those accepted as citizens (with the full set of rights and responsibilities associated with the status) or simply as economic assets, will shape the national community.

Through the years, Canadian immigration policy has been formulated as a tool to construct a national population that is healthy and contributes to the economic growth of the country.[3] Those who do not fit the paradigm have been systematically excluded. The Canadian state has relied on the medical inspection of immigrants in order to halt the influx of undesirable applicants. The screening of immigrants has been based on the assumption that only those who might be of use to Canada should be allowed in.[4] Whereas concerns about individuals with physical or mental disorders were initially related to their negative economic as well as social and moral impacts, the emphasis has now been restricted to economics. Today, persons with diseases, disorders, or disabilities are not excluded because they bring the threat of race degeneration, but because they might constitute an economic burden. This shift aside, Canadian immigration policy has remained one of the main methods of regulation of the national population and labour force. As in many other states within the Western system of capitalism, Canada has relied on a view of individuals as worthy only insofar as they are useful to the material growth of the country. Wendy Brown advances the argument that neo-liberalism, understood not just as a set of economic policies but as political rationality, has resulted in the dismissal of "the fundamentals of liberal democracy"[5] and in their replacement with "market criteria of cost/benefit ratios, efficiency, profitability, and efficacy."[6] Albeit two of the main objectives in the IRPA are facilitating family reunification and fostering respect for human rights,[7] it is increasingly evident that both goals take a back seat when confronted with economic considerations. Throughout this book, I have shown that immigration control through medical assessment has become one of the preferred instruments employed by the neo-liberal state to prevent supposedly unuseful and unprofitable subjects from entering the country.

The trend towards a neo-liberal rationality and governmentality is worrying, especially because in the current economic situation it feeds on the insecurity of the population. Societies in the capitalist world are rendered so afraid of losing their wealth and privileges that they are willing to give up their commitment to democratic principles and rights in exchange for the modest promise of keeping the economy going. Barry Smart observes that an electoral majority has emerged who is willing to support "any political constituency that promises not to disturb or disrupt their immediate comfort and contentment."[8] In the whirlwind of concerns about economic recovery and increased productivity and efficiency, Canadians with disabilities end up at the margins of society while foreigners with disabilities are unceremoniously barred from entering. Canadian society has developed a work culture that grants social status on the basis of people's ability to work, and the definition of work we have become accustomed to fails to acknowledge that persons with disabilities are productive subjects.[9] Further, the concept of productiveness is problematic in itself, and we must categorically reject any argument that human beings are meaningful only if economically productive. This approach limits an understanding of the complexity of human life. In a society that has grown accustomed to the idea that everything, people included, has a price, we need to be reminded that "the most important things are beyond price. How much is your mother worth? Or your sister or your child?"[10]

The discourse surrounding persons with disabilities goes beyond the impact of neo-liberalism within the Canadian political, economic, and societal context. Whereas the obsession with productivity explains a piece of the puzzle, it leaves untouched other important assumptions. Persons with disabilities have been portrayed as the prototype of unproductiveness, even though there is plenty of evidence to indicate that their exclusion from the labour market is a result of barriers created by society rather than the product of disability itself. They have been perceived as costing the system in terms of both services and money without giving back anything substantial, a claim that has been proven untrue. Further, while neo-liberal governmentality has targeted various minority groups that at one point or another were perceived as unproductive, this targeting has shifted depending on the historical context with some groups coming in, others moving out. Persons with disabilities represent the only exception to this rule: they never move out. Why the exception? My analysis indicates that whether we look at the political, societal, or legal discourse, Canada remains uneasy with the idea of having disabled individuals within its borders. My argument throughout these pages has been that this uneasiness results from interlocking concerns around productivity with those of a more ideological nature. Drawing from Kay Anderson's work on Vancouver Chinatown, I argue that in order to understand the situation experienced by immigrants with disabilities in the Canadian context, it is crucial to consider their position within the ideological framework of Canadian society.[11] Whereas economic, social, and political circumstances affect how Canadians react to immigrants, I have maintained that the rejection of medically inadmissible applicants also rests on cultural assumptions about the inherent unworthiness of persons with disabilities, particularly when they also happen to be outsiders rather than nationals. The examination of the public discourse entertained by federal politicians, the press, and the legal system provides evidence to validate this claim.

Devaluation of persons with disabilities is reflected in the marginalization of disabled citizens and in the rejection at the border of medically inadmissible immigrant applicants. Since its inception, Canadian immigration policy has steadfastly refused entrance to those who fell outside the ideal healthy, productive individual. The language in subsequent versions of the Immigration Act and its accompanying regulations has been modified over time in order to make it less offensive to constantly evolving sensibilities, yet its content has remained unaltered. The refusal to acknowledge the non-material and non-economic contributions of potential immigrants has been scarcely explored by immigration scholars, who have traditionally concentrated on issues of integration and conflict among different cultures, thus bypassing the debate around the role that immigrants play within the national body aside from economics. Persons with disabilities are a reminder that difference, understood as, in the words of Iris Marion Young, "unassimilated otherness,"[12] is a value rather than a barrier and that there are many different ways to look at and be in the world. The dismissal of disabled people's contribution impoverishes civil society. Tanya Titchkosky reminds us that "being in the world, disability can teach us about that world."[13] And yes, the world is more than the economy, profit, and productivity. Disability can teach us that despite our illusions of independence, we are all interdependent, and this does not signify failure. Instead, we have the comforting assurance that we are never alone in the world, always in it together.

The discrimination entrenched in article 38(1)(c) of the IRPA stems from the widespread use of the medical model in understanding the lives of those with disabilities. It rests on the assumption that a medical examination is sufficient in assessing the capacity of persons with disabilities to meaningfully participate in society. Exclusion based on impairment or disability denies the right to equality afforded to all human beings by both the Canadian constitution and several international agreements Canada has signed. Ruling on the social, intellectual, and emotional potential of the disabled person to participate in the development of Canadian society, the law disregards other essential factors such as a person's age, ability, and background. In reducing the person to an economic factor, the article removes any social, moral, or cultural contribution of the individual to society's development, and it denies that person's right to become a Canadian citizen exclusively on the grounds of disability. Ann Shearer warns, "the 'contribution' people make to their society is not simply to be defined in narrow economic terms."[14] We all have gifts and talents that enrich society, as well as the right to express them. Paul Hunt observes, "*no* difference between men, however real, unpleasant and disturbing, does away with the right to be treated as fully human."[15] These rights are not something to be earned by competence. Asking individuals to prove their worthiness indicates our inability to recognize the common humanity binding us together. And yet, article 38(1)(c) mandates the doctor in charge of the medical examination, most often a general practitioner rather than a specialist, to assess whether a person with a disease or disability might constitute an excessive burden on health or social services. In this context, the term excessive refers to costs that would likely exceed the average Canadian per capita costs over five years – though in some cases the limit can be extended to ten – as established by Statistics Canada. The reference to the average Canadian is arbitrary at best as it does not account for differences with respect to age, gender, lifestyle, and mental or physical disabilities. A time limit of five to ten years is also problematic when considering the current speed of medical discoveries and new treatments. Furthermore, medical examinations are far from the best way to assess persons with disabilities, since disability is not the same as sickness, though they are often overlapping conditions. It would be a definite improvement for CIC to leave behind the medical perspective, embracing instead a new approach centred on the notion of readaptation and focusing on possible ways to decrease the effects of barriers.

Persons with disabilities must be recognized as an essential part of society – not simply valuable if they can integrate into it. Within disability studies, it is now widely accepted that disability is not a characteristic of the individual body but the result of social practices that devalue difference.[16] Whereas in the past the terms disability and handicap were often used interchangeably, in 1980 the World Health Organization (WHO) adopted what it believed to be a more precise classification of impairments, disabilities, and handicaps. The new terminology was meant to distinguish between disability, which refers to functional limitation of the individual, and handicap, which indicates the loss or limitation of opportunities to take part in the life of the community on an equal level with others – loss and limitation resulting from the encounter between the individual and their surrounding environment.[17] The distinction introduced in 1980, while acknowledging that handicap is a social construction, fell short of appreciating that disability is also socially constructed. It is disheartening that despite its own efforts, the WHO has so far been unable to recognize that its terminology is still validating an understanding of disability as "coterminous to illness" and "a net drain on society."[18] One of the outcomes of such an understanding is the provision of medical admissibility discussed in these pages, which continues to blend together disease, disorder, and disability. This confusion of terminology results in sanctioning a perception of disability as something that is unwanted and in need of elimination – both for the well-being of the individual affected and of society as a whole.

Article 38(1)(c) of the IRPA defines persons with disabilities as medically inadmissible to Canada, thus implying that they are unable to integrate, which situates them in a position of disadvantage with respect to other immigrants. Since the early 1980s, groups such as l'Association multi-ethnique pour l'intégration des personnes handicappée du Québec have asked the federal government to review Canadian immigration law to be consistent with the spirit of the Charter and with international human rights instruments to which Canada is signatory.[19] Despite repeated calls for a thorough review of the law, little has been accomplished in terms of legislation aside from the change introduced in the IRPA that eliminated the admission bar for excessive medical demands for refugees, sponsored spouses, common-law partners, and dependent children.[20] This partial concession does not satisfactorily address the problem of how the Canadian state relates to immigration and disability. Furthermore, independent immigrant applicants with disabilities are still denied access on the grounds their condition would place excessive demand on the system. Additional research is required in order to find an adequate response to developments (or lack thereof) in legislation on the subject of immigration of persons with a disease or disability, especially within a country that presents itself as being committed to equality, protection of minorities, and openness to diversity. It is significant that the state keeps rejecting accusations of unconstitutionality when reminded that the medical admissibility provision appears to be in violation of the spirit, if not of the letter, of the Charter. The rejection suggests that there are exceptional circumstances that allow the state to ignore its own laws and principles, such as its commitment to the protection of minority groups, in the name of a higher good – in this case, the maintenance of an economically viable national body. As Giorgio Agamben points out in *State of Exception,* the state increasingly appeals to the necessity for spaces of exception removed from judicial oversight as a technique of government.[21] In so doing, it creates spaces where the "normative aspect of law can thus be obliterated and contradicted with impunity by a governmental violence that ... nevertheless still claims to be applying the law."[22] In the field of immigration, Canada claims to be applying the law while appearing to be in violation of its own internal and international commitments and values.

In his compelling enquiry, Agamben traces the origins of the state of exception to the emergency situations created during times of war when governments used to suspend the constitution in order to acquire the full powers needed (or so it was argued) to protect the country. Agamben demonstrates that in most cases, military emergency outlasted the conclusion of the hostilities and was transformed into economic emergency in times of peace; both sets of emergencies required the adoption of a state of exception.[23] This turn of events is what seems to be happening in Canada: part of the constitution – section 15 of the Charter – is ignored because, in the government's perspective, it might damage the Canadian economy and, in so doing, imperil the state. Whereas ignoring the law does not equate with suspending it, I argue that the Canadian state is currently ignoring section 15 as an alternative to suspending it. In this case, ignoring the law accomplishes the exclusion of immigrants with disabilities without the need for justification, something that would inevitably be required if section 15 was formally suspended. The rationale provided by the government that the admissibility clause does not discriminate against disability but focuses instead on excessive demand looks spurious. The admissibility clause *does* discriminate against disability, and it does so because of the mistaken and unsubstantiated assumption that persons with disabilities are burdens. As a consequence, preventing the entrance of medically inadmissible subjects becomes a necessity for the Canadian state. And yet, what is necessity? Despite the fact that "writers continue more or less unconsciously to think of [it] as an objective situation," Agamben notes that necessity "entails a subjective judgment, and ... the only circumstances that are necessary and objective are those that are declared to be so."[24] Hence, the necessity to keep undesirable immigrants out is the outcome of a conscious choice within a specific context from the point of view of those holding power. I think that everyone can recognize the arbitrariness in such a situation.

If we accept that state necessities are in fact conscious choices of convenience made by those holding power, we can then explain immigration decisions through the interpretative lenses of the Foucauldian discourse on biopolitics and state racism. Foucault defines the focus of biopolitics as "the population as a political problem."[25] Within this framework, disease and disability come to be understood as "phenomena affecting a population."[26] The state response to this biopolitical problem is different for nationals versus outsiders. In the case of nationals, the state answers by creating a whole apparatus of control made up of not only charitable institutions but also other mechanisms such as "insurance, individual and collective savings, safety measures, and so on."[27] When faced with outsiders, the answer is instead found in state racism, which Foucault describes as "a way of introducing a break into the domain of life that is under power's control: the break between what must live and what must die."[28] In this context, death does not exclusively refer to physical death but includes political death as manifested in the process of expulsion or rejection. When Agamben adopts Foucault's analysis and claims that "Western politics is a biopolitics from the very beginning,"[29] what he means is that the state justifies the exclusion of certain subjects as a political necessity that takes precedence over any other law, including the constitution of the land. Under this logic, immigration rules, among them the medical admissibility provision, are legitimized and cannot be questioned by recourse to the Charter.

Throughout this book, I examined how medical admissibility has been approached in Parliament, the press, and the courts. The analysis of public discourse at the centre of each chapter illustrated that language reveals plenty about the way selected collectivities have framed certain cognitive categories – namely, persons with disabilities and immigrants – and acted to translate such cultural assumptions into polity. Within this context, if we recognize that insofar as they have agency they also have a degree of intentionality, it could be inferred that the different sectors whose public discourse was analysed throughout these pages were operating according to a conscious or subconscious collective intent. Indeed, it would not make sense to talk about the responsibilities shared by parliamentarians, the press, and the courts in devaluating and excluding medically inadmissible immigrants if we do not recognize any intentionality in their discourse. One of my goals has been to move beyond a mere illustration of public discourse and instead reach an understanding of what was conveyed through the spoken and written words adopted by the three selected groups in their discussions.

While chapter 1 provided an overview of the evolving Canadian approaches to disability and immigration, chapter 2 focused on the interventions of federal politicians who passed and maintained the medical admissibility provision. Throughout the chapter, I conducted an analysis of debates and discussions held in the House of Commons and Senate. All parties have historically remained steadfast in their support for the provision of medical admissibility, with only a handful of members openly criticizing it. While the country's immigration policy has traditionally been a subject that has found parliamentarians divided along clear party lines, the policy and associated regulations referring to the entrance of medically inadmissible persons represent the exception where antagonism has been replaced by overall agreement. Politicians' interventions indicate that the main concerns underlying the admission process was whether the applicant was going to become an economically productive member of society or a burden. Federal politicians have adopted a utilitarian approach to immigration, thus letting in only those considered profitable investments. Since its formulation, Canadian immigration policy has had a manpower orientation that has remained unaltered during the last century. Immigration doors were opened and closed in accordance with economic fluctuations, yet the basic assumption that immigrants should serve the interests of the country remained unquestioned. This stance validates the argument that the state does not act as a benevolent entity or, in Adelman's terms, as a charitable institution;[30] on the contrary, it operates exclusively in its own interests. In selecting immigrants, the state is unconcerned with presenting itself as moral or virtuous – it instead aims at defining its national body according to considerations of usefulness.

Despite this continuity, a change took place among politicians starting in the mid-1950s due to the moderate advancements made in the legislation as a result of pressures exercised by better organized advocacy groups. The 1950s witnessed the passage of several pieces of legislation advancing the status of Canadians with disabilities. These improvements did not translate in gains for disabled immigrant applicants as the medical admissibility provision was untouched. Nevertheless, they reflected a new perception of disability. In this period, the number of parliamentary interventions regarding immigrant applicants with a disease or disability drastically dropped, and most of the modifications to the legislation were primarily implemented in the form of regulations put in place by the government or senior officers working in the department responsible for immigration. Even when decisions were made in Parliament, the language was intentionally left vague so that the provisions would have been next to meaningless if not accompanied by regulations prepared elsewhere. The issue of medically inadmissible immigrants was becoming contested in light of the new emphasis given to the rights of minority groups within Canadian borders. Politicians' reluctance to talk about the subject translated in its disappearance from parliamentary debates. It was preferable to stay away from discussions on how to keep out foreigners with disabilities when declaring with pride that Canada was eager to integrate its own disabled population.

The re-emergence of a lively discussion was instigated by the passage of the Charter of Rights and Freedoms. In conjunction with the tireless efforts of several disability activists up to as well as after 1982, the Charter contributed to "the shift from disability as a charity concept to legitimizing disability as a status entitled to rights."[31] The equality clause provided minority groups with a tool for having their rights recognized. For the first time, Ottawa could be brought to court to defend its legislation. This compelled politicians to revisit and change existing laws, thus bringing them in accordance with the Charter. Faced with the possibility of a law being deemed unconstitutional, the political class was forced to reappropriate a role too often delegated, largely for reasons of convenience, to the bureaucracy. Unelected civil servants undoubtedly exercised a strong influence on the way immigration policies were formulated and implemented, yet I have not discussed their role as the ultimate responsibility in passing legislation rests with politicians. It is not without reason that the courts, when refusing to intervene, always refer to Parliament as interlocutor and the body responsible for the legislation on matters of immigration. Following the passage of the Charter, the number of parliamentary interventions contesting the inadmissibility of certain categories of immigrants on the grounds that it violated section 15(1) of the Charter increased. Unfortunately, while the discussion has been revived, significant changes have failed to materialize. Governments – Liberal and Conservative alike – have fought hard, mainly through issuance of ministerial permits, to prevent people from going to court to assess the constitutionality of the medical admissibility provision. Until the legislation is questioned in court, it is relatively easy to divert the attention of the public to more pressing matters, effectively maintaining the status quo.

Another issue emerging from parliamentary debates is the role of the state within Canadian society. Aside from a few individual interventions, parliamentarians acted to protect what they believed were the interests of the country. They presented themselves as neither heroes nor villains, but officials with a specific mandate requiring them to ignore moral considerations in order to prevent the country from crumbling under perceived external pressures. It seems to have been their belief that the state was not supposed to be ethical in its acting. And yet, that is beyond the point. I do not expect politicians to conform to ethical standards. I recognize that the creation of any ethical state through the actions and discourses of its political representatives would be highly questionable: Whose ethics are being brought forward? Who decides which measures are ethical? Moral criteria are culture specific and vary across groups. They imply a value judgment and therefore bring back distinctions between right and wrong, what/who is acceptable, and what/who must be rejected. Daniel Bensaïd warns us that ideas such as Truth and Good belong to the field of religion; when they intrude into the political arena, the state risks transforming itself into a theocracy.[32] My call here is not for an ethical state, which is neither needed nor desirable. At the same time, political figures acting exclusively under the premises of preservation and self-interest are in danger of overlooking the value of human life and ignoring that individuals have the right to be treated with dignity. I wonder whether no ethics at all is actually possible or whether this would simply result in a different kind of ethics based on inaction. I suggested elsewhere adoption of the concept of human rights as a way out of the impasse implicit in the dualism of ethics/no ethics.[33] The concept of human rights brings to the forefront the idea of "unassimilated otherness,"[34] diversity without rejection, and inclusivity rather than discrimination. Valorization of human life is something that goes beyond ethics, and it brings us back to "the universality of some basic premises ... which constitute a constant link among people."[35] Whereas the notion of human rights is less legally binding than the one of citizenship,[36] it represents a better choice exactly because of its indeterminateness and flexibility.

Moving from Parliament to the editorial room, chapters 3 and 4 investigate the public discourse animating the press. Articles, editorials, and letters that appeared in the *Toronto Star* and the *Globe and Mail* indicate that both papers have traditionally been in favour of banning those perceived as endangering the economic well-being of the country. The picture emerging from the investigation of the two papers is at best one of uneasiness with respect to foreigners likely to become a burden. However, this approach tended to change when the discourse switched from general considerations to distinct cases. Each time the papers reported specific situations where individuals were rejected at the border or deported from the country, the readership and contributors overwhelmingly sided with the unfortunate victim against heartless politicians applying a heartless law. Showing compassion for an exceptional case was well within the boundaries of what was perceived as acceptable. Generosity towards a particular individual looked good, yet, at the same time, it did not force reconsideration of the status quo. The exception indeed validated the norm. Family unity also provoked sympathy for it seemed unfair when families were separated because one member was deemed medically inadmissible. This sympathetic attitude is in tune with Canada's long-time commitment to facilitate family reunification and supports the emphasis Canada puts on the family as a primary social institution.[37]

Despite the sustained support to the government shown in both papers, one main exemption and two subsequent changes are worthy of mention. The exemption to the endorsement of the medical admissibility provision spans the entire period investigated. A number of readers and members of the press, while willing to keep medically inadmissible foreigners out of their country, did not react positively when their conationals occasionally received the same treatment abroad – particularly when they were Canadians who were refused access to the United States. Regulations adopted with *others* apparently did not fit well when applied to *our own*. Such a state of affairs seems to confirm Arendt's argument that aside from the rhetoric around human rights, only citizens can claim "a right to have rights."[38] Today more than ever, citizenship is "a contestable notion"[39] whose meaning is far from unanimously accepted, and yet it remains that status alone that gives recognition to the individual. It is my opinion that the Charter might have the potential to invalidate such argument, though it has not yet succeeded in doing so. I am confident it will succeed the moment civil society starts pressuring the Canadian state and judicial system to recognize that the Charter's mandate has been intentionally ignored for political reasons of convenience to the detriment of individuals who should have been protected from discrimination under section 15. As shown throughout this book, little seems to happen until civil society gets outraged to the point of demanding that the status quo be altered.

The two aforementioned changes occurred instead in the years following the Second World War and after the passage of the Charter in 1982. Both changes found an echo in Parliament where the debate was affected by shifting attitudes within society at large, mainly as a result of the indefatigable work of disability organizations and activists. Soon after the war, although the policy of medical admissibility failed to undergo any significant adjustment, Canadian society became more receptive to a discourse of respect for, and integration of, persons with disabilities, partly because so many of its own citizens had come back from the front lines as permanently or temporarily disabled. In 1945 Canada also joined the UN and began developing the reputation of being an accepting and compassionate country. Whether that corresponded to the truth or was just a myth, Canadians were proud of this perception and were determined to translate it into reality – at least as much as required to believe it was true. Hence, they were willing to show greater flexibility with the rules governing admissibility of disabled immigrants. I am inclined to see these modifications as dictated by reasons of convenience rather than the outcome of questioning the basic assumptions around the place of disability within society. Nevertheless, I do not exclude it might sooner or later translate in real change in mainstream understandings of disability. While I am not ready to go as far as to argue that change is on the way, I remain hopeful.

For the period following the passage of the Charter, it is evident that the document has impacted the way Canadian society approaches issues of equality and rights for individuals belonging to minority groups. The importance the Charter has acquired in the last three decades for Canadians as a symbol of their identity cannot be underestimated. In a relatively short period of time, the Charter has become the concretization of what Canadians stand for – or would like to. Although various sectors of society have remained essentially supportive of the medical admissibility provision, the Charter has given legitimacy to the claims of several groups and organizations working to advance minority rights and has strengthened their determination. In the long run, the impact of the Charter on Canadians' perceptions has affected the press coverage of medically inadmissible applicants. The Charter has raised questions among the public on issues that were before quite comfortably ignored. More work is required, but the debate that has been generated is promising.

In chapter 5, I reviewed several court cases dealing with medical admissibility. The discussion illustrated that there has been no ideological shift in the state approach to the issue, despite changes in the language registered in the discourses of politicians and newspapers. The time frame covered in this section was restricted to the years after the passage of the Charter since the document provided an incentive for claimants to contest the provision in the legal arena. However, only one of the seventeen litigations considered made direct reference to the Charter. In all the remaining lawsuits, claimants contested the way the article was applied to their specific circumstances rather than its content. This limitation could be explained by the fact that the defendant, the Department of Citizenship and Immigration, maintained that the medical admissibility clause was not centred on the notion of disease or disability and therefore was neither discriminatory nor unconstitutional. According to the defendant, the focus of the clause was excessive demand. As a consequence, a majority of the claimants opted to leave aside the issue of unconstitutionality in order to avoid a process that would have been expensive, time-consuming, and likely unsuccessful in light of the renowned conservatism of the courts. There are different constraints preventing people from fighting for their rights in a court of law. This variance clearly invalidates the common assumption that the law is equal for everyone. Who you are, your position in society, your ability to publicize your case, and your resources all matter in deciding whether a claim is going to be successful.

The legal action undertaken in the late 1990s by Angela Chesters was the only exception among the seventeen court cases considered. Chesters challenged the medical admissibility provision as unconstitutional for discriminating against persons with disabilities, thus violating section 15 of the Charter. Despite its failure, Mrs Chesters's attempt represents an important moment in the journey towards a more consistent approach to equality in the Canadian justice system and society. Whereas the court refused to recognize that there are still laws in violation of the Canadian constitution, the hearing ignited a debate within the public and this upheaval was undoubtedly a major achievement. Although the courts have not always held a progressive interpretation of the Charter, the mere possibility of recurring to the legal sphere has empowered minority groups, the disabled community being one among several beneficiaries. In spite of the progress made, more needs to be done in pressuring the courts to finally address the apparent unconstitutionality of Canadian immigration law. So far, no valid justification has been provided by the state for maintaining a practice of discrimination that has shattered so many lives.

If we look at the different chapters as pieces of the same puzzle, there are a few interesting resonances that connect the discourse codeveloped by the three selected institutions analysed throughout the book. Overall, it is possible to identify three separate (although at times overlapping) historical phases. The first phase covers the years from 1902 until the first half of the 1940s. Throughout this period, rejection of potential immigrants with diseases or disabilities was overwhelmingly motivated by the two major concerns of productivity and sanity. Canadian politicians and the press were in agreement that accepting a diseased or disabled immigrant was going to negatively affect Canada economically as well as corrupt its genetic make-up. It should not be forgotten that, throughout this time, eugenics was extremely popular not only among members of the scientific community but also in the population at large. The coverage provided in the *Toronto Star* and the *Globe and Mail* makes evident that the dominant discourse regarding the landing of defective subjects was unquestionably accepted. If anything, the press put pressure on the government to tighten the restrictions preventing undesirables from entering the country. Concerns were also motivated by considerations of foreign policy, particularly with respect to the reception that Canadian policy was having in the country's two major partners: the United States (mainly because of economic ties) and Britain (due to economic, cultural, and political connections). Whereas few heartbreaking stories were reported, such as the one of Margaret McConachie, and generated some sympathy among readers, the overall approach to immigration of undesirables tended towards a refusal to let them in.

The second phase covers the period from the end of the Second World War up to the late 1970s–early 1980s. After the horror of the Nazi experience, support for eugenics substantially decreased, as did concerns around sanity in the selection of immigrants. The reasoning embraced by federal politicians for continuing to exclude foreigners with diseases or disabilities from Canada became more and more centred on economic considerations, particularly their supposed unproductivity and therefore their uselessness to Canada. At a time when the country needed workers, it made little sense to accept individuals who were assumed to be unemployable. A secondary reason that started to gain validity, especially after the passage of the Medical Services Act in the late 1960s, was the potential cost these subjects could create for Canadian taxpayers. It is not surprising that while legislation was being passed to support and provide benefits to disabled Canadians, an almost opposite approach was taken in the case of disabled or diseased foreigners. Throughout this time, the press maintained its support for a policy of exclusion, although it gradually increased its reporting of individual stories (largely because of their selling potential) and the request for selected exceptions, especially in cases that saw families separated. When pressures from the press and society at large became overwhelming, the federal government agreed to bend the rules provided the overall framework remained unaltered. The 1970s witnessed indeed a growing recourse to ministerial permits for humanitarian and compassionate reasons. Some of the language in the legislation was also softened to better suit the sensibilities of the Canadian population.

The third and last phase covers the years from the passage of the Charter until the end of the investigation in 2002. The focus on employability gradually lost steam as a justification for the admissibility provision as it became apparent that it was an unresolvable contradiction to claim disabled persons should have been integrated in the labour force but only as long as they were Canadians. Obviously, the ability of an individual to work is not affected by citizenship status. The policy was therefore adjusted to emphasize the potential costs immigrants with diseases or disabilities could generate for Canada. The novelty characterizing this period is represented by the passage of the Charter of Rights and Freedoms, which affected the discourse formulated in the press and in the legal arena. Concerning the press, the Charter's influence in shaping Canadians' perceptions and attitudes was exemplified in a more critical approach to the medical admissibility provision. While it would be incorrect to argue that currently the press is opposed to the provision, I think it is a fair statement that in most cases the press is unwilling to endorse government decisions barring disabled or diseased foreigners. Within the legal arena, although the courts have failed to recognize the discrimination of the medical admissibility provision against one of the minority groups under the protection of the Charter, the document has provided new impetus for those physically in Canada to seek redress in a court of law.

In writing this book, my hope was to contribute to a better understanding of the underlying assumptions present in the public discourse around Canadian immigration policy from the beginning of the twentieth to the early twenty-first century. I tried to show the place immigrants have been assigned through time in the life of the country and how this has affected their reception. I also argued that Parliament, the press, and the court system, in their roles as elites with unique access to public discourse, are not, despite their claims to the contrary, three separate domains, each one of them proceeding with its own rules and at its own time. They are instead inherently connected and, as shown throughout the different chapters, each step taken in one setting is reflected in the others. As argued by Ferri and Connor, "these lines of power often combine, overlap, and intersect with one another."[40] The tripartite structure adopted throughout the book has hopefully allowed readers to witness first-hand these intersections and overlaps.

Above all, I am confident that this book proves that the state tends to maintain the status quo whenever possible, taking action only when influenced by the changing attitudes of politicians, the press, and the court system, which are in turn pressured by civil society, particularly activist groups. All of the changes the medical admissibility provision underwent in the hundred years considered, as small as they might have been, were ultimately made possible through the pressures exercised by sectors of civil society and their dissatisfaction with the existing situation. In lieu of what is exposed in the previous chapters, it is indeed unreasonable to assume that certain discussions on the inappropriateness of the language used in the legislation, as well as several particular provisions, such as the sponsorship of disabled spouses and dependants, would have reached the floor of Parliament if not for politicians recognizing these issues were acquiring some resonance with their constituency. This realization came through a number of channels, among them petitions, individual letters, and press coverage. Also decisive in affecting politicians' responses and stirring up state reaction – the two are not one and the same and yet are intrinsically linked – were several court proceedings conducted in those years. The fact that legal avenues were pursued by those who had been rejected from Canada compelled the state to concede something in order to avoid the cost involved in a trial. Even worse was the risk of having the courts take a bold step and mandate overhauling the legislation. Nonetheless, coming into the game of the legal system was the consequence of an act of Parliament through the passage of the Charter of Rights and Freedoms in 1982 and the inclusion of persons with disabilities among the groups protected in its equality clause. The inclusion was made possible, in turn, via pressures exercised by activist groups such as the Coalition of Provincial Organizations of the Handicapped, the Canadian Association for the Mentally Retarded, and the Canadian National Institute for the Blind. It is therefore important to consider the pivotal role played by a number of these advocacy organizations. Their contribution was significant in at least three different ways: educating the general society on issues of disability and equal opportunities, pressuring politicians through written submissions and direct interventions in various parliamentary committees, and supporting those who chose to legally dispute government decisions, either financially or through their role as intervener in court cases. All things considered, there are several reasons to be optimistic. The Charter has provided and will certainly provide in the future opportunities, both within and outside the legal system, for challenging the current legislation, which will create a more equitable society. Different organizations and advocacy groups will continue to impact the way the country perceives persons with diseases or disabilities. As Wood and Isin remind us, identity is "a process of becoming rather than being."[41] I hope Canadian identity will find its way of becoming more inclusive and open to diversity.

This book does not represent a conclusion. Several questions remain unanswered and should be further explored. First, it would be worthwhile to examine the reasons behind the traditionally scarce involvement of various immigrant organizations in contesting the medical admissibility provision. To this day, the main actors advocating for a less discriminatory approach to medical admissibility have been Canadian organizations advocating on behalf of persons with disabilities. They have been pressuring Parliament and the courts while also carrying forward a work of advocacy and education within the society at large. This process has however been a difficult one as the state has exploited the tensions existing between Canadian and non-Canadian individuals with disabilities. It has endeavoured to create a divide between the two groups by arguing that acceptance of medically inadmissible immigrants would result in economic expenses, to the detriment of disabled Canadians. This discrepancy was made clear in the defendant's argument in *Chesters v. Canada*, when CIC maintained that the rejection of Angela Chesters had nothing to do with disability but was exclusively motivated by economics and the intent to protect Canadian health and social services against excessive demand so that Canadians could better access them. Whereas some individuals, as revealed in several letters sent to newspapers, fell victim to such rhetoric, disability organizations have largely escaped the flawed logic of these arguments. They have pointed out that rejection of foreigners with diseases or disabilities impacts Canadians insofar as it sends a message implying who is valued and who is considered as a burden within the national body. Immigrant organizations have instead been missing from the debate. This absence could be a result of difficulties in creating a common front among groups that are ethnically, linguistically, and culturally diverse. It would be helpful to conduct a more in-depth analysis on the matter.

Further research could also explore the intersection of race, ethnicity, class, and gender in the process of immigrant selection and how they combine to determine whether immigrant applicants with disabilities are excluded or allowed into the country. Chapter 5 has, for instance, revealed that access to the legal system is restricted to those who can pay for it. Equally important, since the Supreme Court decision in *Hilewitz v. Canada* and *De Jong v. Canada*, the ability to pay for social services has become one of the criteria for assessing medically inadmissible immigrant applicants. On issues of race and ethnicity, chapters 3 and 4 have shown that readers and journalists had different reactions depending on whether the person rejected was a white, middle class, educated immigrant like Angela Chesters or an outsider such as Mrs Kumar, an Indian woman who was deported from Canada in 1980 after being diagnosed with mental illness. Further investigation should focus on issues of gender: looking at the coverage provided in the *Toronto Star* and the *Globe and Mail*, it becomes apparent that, on the whole and especially in the years before the 1980s, a large number of rejected or deported persons were women. Without an analysis of the actual data for immigrants denied status throughout the entire period, it is impossible to make credible conclusions. Was the percentage of women over men significant? If so, a gender analysis would help in understanding the reasons behind statistical variations. Further, it should be considered that, especially in earlier years, the number of women applying under the independent immigrant category was lower than at present, and most of those denied admission were likely applying as sponsored immigrants. It is conceivable therefore that more women than men were receiving newspaper coverage simply because these women were members of a family unit and, as we have seen, cases of family separation were more likely to find a sympathetic audience.

Another element requiring additional attention is the impact of the current immigration policy on immigrants with a disease or disability who, for different reasons, have been allowed to enter the country. How has their experience impacted on their participation in Canadian society? To what degree do they feel like citizens in Canada? In a preliminary research project conducted in 1998, Judith Sandys pointed out that several immigrants interviewed reported being reluctant to access the service system in Canada. They repeatedly mentioned that their experience of the immigration process had been highly stressful and left them with a sense of vulnerability, which discouraged them from seeking access to services.[42] It would be interesting to investigate whether these immigrants have experienced the same difficulties in other aspects related to citizenship. How does the perception of being undesirable created through the immigration process affect the way these individuals act as citizens of the country? Does it discourage them to reclaim their citizenship and its associated rights? Does it silence and marginalize them?

Then again, what role is left for the Charter in advancing the rights of foreigners with disabilities in Canadian society? Will the traditionally conservative approach shown by the courts change in the foreseeable future? I believe that the full potential of the Charter has not been adequately explored. Whereas in 2002, in the case of Angela Chesters, the court refused to address any of the criticisms presented by the intervener at the hearing, the CCD, it is unlikely this attitude will remain successful in the long run. The judicial system does not exist in a vacuum but is affected by and affects the society it is part of. New approaches towards persons with disabilities are already impacting the way society relates to this minority group. The courts will be unable to ignore these changes forever and eventually will have to come to terms with them. It remains to be seen when, how, and to what degree this negotiation will occur.

While the judicial system cannot be separated from the society around it, the same holds true for Parliament. Pressures from the public and the courts will eventually force politicians to review the legislation on medical admissibility, even though Parliament has usually preferred to procrastinate on any change or revision. As seen in chapter 2, politicians have traditionally been quite receptive to demands coming from the broader society since they are aware that ignoring such calls could potentially cost votes. One of the first aspects of the legislation to be revised will likely be the widening of the exemption to the medical admissibility provision for members of the family class other than immediate dependants. Considering that recent census figures indicate immigrants as the main source of population growth all around the country, it is fair to assume that in time immigrants will exert pressure for a relaxation of the medical admissibility provision since the percentage of individuals with a disease or disability among older family members is usually higher than among independent workers.

Last but not least, it would be helpful to look at the international stage and compare how different countries have approached medical admissibility. Although there have been a number of comparative studies related to the immigration policies in countries of immigration (United States, UK, Canada, and Australia), other areas of the world have not received similar attention. It would be worthwhile to focus on countries that have historically never experienced a steady influx of immigrants but are now feeling the pressure. Some of the countries in the European Union (EU) could be an obvious starting point. Among others, Italy is increasingly experiencing massive immigration from Eastern Europe, Northern Africa, and the Middle East at a time when unemployment is high and the immigration policy of the country is inadequate to handle the arrivals. The headlines on major national newspapers are increasingly emphasizing issues such as crime and illegal activities carried out by foreigners. This attention is something Canada knows all too well from its own past. Italy has a welfare system much like Canada, but it does not have a medical admissibility provision in its immigration policy. What will happen when Italians start complaining that newcomers are becoming a burden? What of other countries experiencing a similar situation, either in the EU or elsewhere? These are just some of the unanswered questions on the topic of medically inadmissible immigrants both within and outside Canada that scholars, activists, policy analysts, and elected representatives should explore.

I began this book with my personal story, so it seems fair to conclude it by coming back to the personal. Writing these pages has helped me come to terms with my own process of admission to Canada – the disappointments and the anger of being defined as a burden. I am now a Canadian citizen, and yet I have neither forgotten nor forgiven this country for what it forced me to go through. I am not sure I ever will. But today I can use my experience to denounce a system that I consider profoundly unfair and discriminatory. The above pages are a testament to the courage and resilience of many individuals I came across throughout my research. Their stories gave me strength and support when I felt vulnerable and alone. It's to them that I dedicate this work. Because Canadians should not be allowed to ignore that Vivienne Anderson, Angela Chesters, Margaret McConachie, Herman George Aviles, Harjit Kumar, and many others are not just names written on a paper but actual lives discarded by an immigration law that still exists and must be repealed. Because there are no rights and freedoms if we choose to apply them selectively. Because no one should ever be called a burden the way we were.

APPENDIX: CHANGES TO THE MEDICAL ADMISSIBILITY PROVISION IN CANADIAN IMMIGRATION POLICY, 1869–2001

The following is a list of the most significant changes to the medical admissibility provision contained in Canadian immigration policy that are mentioned in the book for the period from 1869 to 2001. Some of these changes were discussed in the previous pages when they were at the level of bills debated in Parliament. For identification purposes, I have therefore added the bills' names before they were passed into laws. These laws are now accessible in the Statutes of Canada (S.C.). To make it easier for the reader to distinguish between the original acts and amendments to the acts, the former are in bold while the latter are in italics.

The Immigration Act, S.C. 1869, c.10, s.11(2), and s.12. (One of the main goals of the act was to prevent disease from entering Canada. Although there were few actual restrictions preventing entrance into the country, diseased or disabled immigrants were to be recorded on a list and required to pay a bond.)

> 11(2). If, on examination, there is found among such Passengers, any Lunatic, Idiotic, Deaf and Dumb, Blind or Infirm Person, not belonging to any Immigrant family, and such person is, in the opinion of the Medical Superintendent, likely to become permanently a public charge, the Medical Superintendent shall forthwith report the same officially to the Collector of Customs at the Port at which the Vessel is to be entered, who shall (except in the cases in which it is hereinafter provided that such bond may be dispensed with) require the Master of the Vessel, in addition to the duty payable for the Passengers generally, to execute, jointly and severally with two sufficient sureties, a Bond to Her Majesty in the sum of three hundred dollars for every such Passenger so specially reported, conditioned to indemnify and save harmless the Government of Canada or of any Province in Canada, or any Municipality, Village, City, Town or County, or Charitable institution within the same, from any expense or charge to be incurred within three years from the execution of the Bond, for the maintenance and support of any such Passenger;
>
> 12. The proper Agent for Immigration may, with the consent of the Minister of Agriculture, make arrangements with the Master, Owner or Charterer of the vessel carrying the lunatic, idiotic, deaf and dumb, blind or infirm person with respect to whom a bond

has been given, or money paid in lieu thereof, or with the Master, Owner or Charterer of any other vessel, for the reconveyance of such person to the port from which he was carried to Canada.

An Act to Amend the Immigration Act, S.C. 1902, c.14, s.1 (also referred to in chapter 2 as Bill 112).

c. *The Immigration Act*, chapter 65 of the Revised Statutes, is amended by inserting the following section immediately after section 24:

24A. The Governor General may, by proclamation or order, whichever he considers most expedient, and whenever he deems it necessary, prohibit the landing in Canada of any immigrant or other passenger who is suffering from any loathsome, dangerous or infectious disease or malady, whether such immigrant intends to settle in Canada, or only intends to pass through Canada to settle in some other country.

The Immigration Act, S.C. 1906, c.19, s.26, s.27, and s.28 (also referred to in chapter 2 as Bill 170). (The act expanded the list of inadmissible immigrants by including those who had been former inmates of mental hospitals or jails; it also added a clause for deportation of those who had become public charges after their landing.)

26. No immigrant shall be permitted to land in Canada, who is feeble-minded, an idiot, or an epileptic, or who is insane, or has had an attack of insanity within five years; nor shall any immigrant be so landed who is deaf and dumb, or dumb, blind or infirm, unless he belongs to a family who accompany him or are already in Canada and who give security, satisfactory to the Minister, and in conformity with the regulations in that behalf, if any, for his permanent support if admitted into Canada.

27. No immigrant shall be permitted to land in Canada who is afflicted with a loathsome disease or with a disease which is contagious or infectious and which may become dangerous to the public health or widely disseminated, whether such immigrant intends to settle in Canada or only to pass through Canada to settle in some other country; but if such disease is one which is curable within a reasonably short time the immigrant suffering therefrom may, subject to the regulations in that behalf, if any, be permitted to remain on board where hospital facilities do not exist on shore, or to leave the vessel for medical treatment, under such regulations as may be made by the Minister.

28. No immigrant shall be permitted to land in Canada who is a pauper, or destitute, a professional beggar, or vagrant, or who is likely to become a public charge; and any person landed in Canada who, within two years thereafter, has become a charge upon the public funds, whether municipal, provincial, or federal, or an inmate of or a charge upon any charitable institution, may be deported and returned to the port or place whence such immigrant came or sailed for Canada.

An Act to Amend the Immigration Act, S.C. 1907, c.19, s.2 (also referred to in chapter 2 as Bill 143).

2. Section 33 of *The Immigration Act*, chapter 93 of the Revised Statutes, 1906, is repealed and the following is substituted thereof:
   33. Whenever in Canada an immigrant has, within two years of his landing in Canada, become a public charge, or an inmate of a penitentiary, jail, prison, or hospital or other charitable institution, it shall be the duty of the clerk or secretary of the municipality to forthwith notify the Minister, giving full particulars.
   2. On receipt of such information the Minister may, in his discretion, after investigating the facts, order the deportation of such immigrant at the cost and charges of such immigrant if he is able to pay, and if not then at the cost of the municipality wherein he has last been regularly resident, if so ordered by the Minister, and if he is a vagrant or tramp, or there is no such municipality, then at the cost of the Department of the Interior.

The Immigration Act, S.C. 1910, c.27, s.3(a), (b), (c) (also referred to in chapter 2 as Bill 102). (A measure was also added requiring all immigrants to prove they were not destitute by showing they possessed at least twenty-five dollars upon landing.)

3. No immigrant, passenger, or other person, unless he is a Canadian citizen, or has Canadian domicile, shall be permitted to land in Canada, or in case of having landed in or entered Canada shall be permitted to remain therein, who belongs to any of the following classes, hereinafter called "prohibited classes", –
   (a) idiots, imbeciles, feeble-minded persons, epileptics, insane persons, and persons who have been insane within five years previous;
   (b) persons afflicted with any loathsome disease, or with a disease which is contagious or infectious, or which may become

dangerous to the public health, whether such persons intend to settle in Canada or only to pass through Canada in transit to some other country: Provided that if such disease is one which is curable within a reasonably short time, such persons may, subject to the regulations in that behalf, if any, be permitted to remain on board ship if hospital facilities do not exist on shore, or to leave ship for medical treatment;

(c) immigrants who are dumb, blind, or otherwise physically defective, unless in the opinion of a Board of Inquiry or officer acting as such they have sufficient money, or have such profession, occupation, trade, employment or other legitimate mode of earning a living that they are not liable to become a public charge or unless they belong to a family accompanying them or already in Canada and which gives security satisfactory to the Minister against such immigrants becoming a public charge.

An Act to Amend the Immigration Act, S.C. 1919, c.25, s.3(1), (2), (3) and (6) (j), (k), (l), and (m) (also referred to in chapter 2 as Bill 52).

3. (1) Subsection one of section three of the said Act is amended by inserting the word "enter or" between the words "to" and "land" in the third line thereof.

(2) Paragraph (a) of section three is hereby amended by striking out the words "within five years previous" in the second and third lines of said paragraph and inserting in lieu thereof the words "at any time previously."

(3) Paragraph (b) of section three is hereby amended by inserting the following words between the word "afflicted" and the word "with" in the first line thereof:- "with tuberculosis in any form or."

(6) Section three of the said Act is further amended by adding the following paragraphs thereto: –

(j) Persons who in the opinion of the Board of Inquiry or the officer in charge at any port of entry are likely to become a public charge;

(k) Persons of constitutional psychopathic inferiority;

(l) Persons with chronic alcoholism;

(m) Persons not included within any of the foregoing prohibited classes, who upon examination by a medical officer are certified as being mentally or physically defective to such a degree as to affect their ability to earn a living.

The Immigration Act, S.C. 1927, c.93, s.3 (a), (b), (c), (j), (k), (l), and (m)

3. No immigrant, passenger, or other person, unless he is a Canadian citizen, or has Canadian domicile, shall be permitted to enter or land in Canada, or in case of having landed in or entered Canada shall be permitted to remain therein, who belongs to any of the following classes, hereinafter called "prohibited classes": –
   (a) Idiots, imbeciles, feeble-minded persons, epileptics, insane persons, and persons who have been insane at any time previously;
   (b) Persons afflicted with tuberculosis in any form or with any loathsome disease, or with a disease which is contagious or infectious, or which may become dangerous to the public health, whether such persons intend to settle in Canada or only to pass through Canada in transit to some other country: Provided that if such disease is one which is curable within a reasonably short time, such persons may, subject to regulations in that behalf, if any, be permitted to remain on board ship if hospital facilities do not exist on shore, or to leave the ship for medical treatment;
   (c) Immigrants who are dumb, blind, or otherwise physically defective, unless in the opinion of a Board of Inquiry or officer acting as such they have sufficient money, or have such profession, occupation, trade, employment or other legitimate mode of earning a living that they are not liable to become a public charge or unless they have a family accompanying them or already in Canada and which gives security satisfactory to the Minister against such immigrants becoming a public charge;
   (j) Persons who in the opinion of the Board of Inquiry or the officer in charge at any port of entry are likely to become a public charge;
   (k) Persons of constitutional psychopathic inferiority;
   (l) Persons with chronic alcoholism;
   (m) Persons not included within any of the foregoing prohibited classes, who upon examination by a medical officer are certified as being mentally or physically defective to such a degree as to affect their ability to earn a living.

The Immigration Act, S.C. 1952, c.42, s.5 (a), (b), and (c). (The act made a clear distinction between preferred classes – British, French, and Americans – and undesirable classes – diseased or disabled subjects, independent Asian applicants, homosexuals, and prostitutes. Although refraining from explicitly discriminating against specific nationalities, the act allowed the cabinet or special investigating officers to refuse entrance to those considered unsuitable to Canadian climate and culture.)

5. No person, other than a person referred to in subsection two of section seven, shall be admitted to Canada if he is a member of any of the following classes of persons:
   (a) persons who
      (i) are idiots, imbeciles or morons,
      (ii) are insane or, if immigrants, have been insane at any time,
      (iii) have constitutional psychopathic personalities, or
      (iv) if immigrants, are afflicted with epilepsy;
   (b) persons afflicted with tuberculosis in any form, trachoma or any contagious or infectious disease or with any disease that may become dangerous to the public health, but, if such disease is one that is curable within a reasonably short time, the afflicted persons may be allowed, subject to any regulations that may be made in that behalf, to come into Canada for treatment;
   (c) immigrants who are dumb, blind or otherwise physically defective, unless
      (i) they have sufficient means of support or such profession, trade, occupation, employment or other legitimate mode of earning a living that they are not likely to become public charges, or
      (ii) they are members of a family accompanying them or already in Canada and the family gives satisfactory security against such immigrants becoming public charges.

An Act to Amend the Immigration Act, S.C. 1967–1968, c.37, s.1 (also referred to in chapter 2 as Bill C-30). (The Point System was introduced.)

1. Subparagraph (ii) of paragraph (a) of section 5 of the *Immigration Act* is repealed and the following substituted thereof:
   (ii) are insane or, if immigrants, have been insane at any time, except an immigrant whose admission to Canada is authorized by the Governor in Council upon evidence satisfactory to him, which shall include the evidence of a qualified medical practitioner, that
      (A) For at least seven years immediately preceding the date of his application for admission, he has neither been a patient in any hospital for the treatment of his insanity

nor suffered any significant recurrence of the symptoms thereof, and

(B) The symptoms of his insanity are unlikely to recur,"

The Immigration Act, S.C. 1976–1977, c.52, s.19 (1)(a) (also referred to in chapter 2 as Bill C-24). (While eliminating formal racial discrimination, the act continued to exclude people who were perceived as likely to become a burden on health or social services. The emphasis was on economics as proven by the introduction, in the 1980s, of the Business class, which consisted of immigrants who were bringing into the country their entrepreneurial talent or investor funds.)

19. (1) No person shall be granted admission if he is a member of any of the following classes:

(a) persons who are suffering from any disease, disorder, disability or other health impairment as a result of the nature, severity or probable duration of which, in the opinion of a medical officer concurred in by at least one other medical officer,

(i) they are or are likely to be a danger to public health or to public safety, or

(ii) their admission would cause or might reasonably be expected to cause excessive demands on health or social services.

An Act to Amend the Immigration Act and other Act in consequence thereof, S.C. 1992, c.49, s.11 (1)(a) (also referred to in Chapter 2 as Bill C-86).

11. (1) Paragraph 19(1)(a) to (c) of the said act are repealed and the following substituted therefor:

(a) persons who, in the opinion of a medical officer concurred in by at least one other medical officer, are persons

(i) who, for medical reasons, are or are likely to be a danger to public health or to public safety, or

(ii) whose admission would cause or might reasonably be expected to cause excessive demands, within the meaning assigned to that expression by the regulations, on health or prescribed social services;

The Immigration and Refugee Protection Act, S.C. 2001, c.27, s.38 (1) and (2). (The act was meant to prevent criminal elements from entering the country, tighten the requirements needed to be accepted as a refugee, broaden skill and training requirements, and assign to same-sex or common-law relationships the same rights of married couples for immigration purposes. Still, no significant change appears in the way applicants with diseases or disabilities were assessed.)

38. (1) A foreign national is inadmissible on health grounds if their health condition
(a) is likely to be a danger to public health;
(b) is likely to be a danger to public safety; or
(c) might reasonably be expected to cause excessive demand on health or social services.

(2). Paragraph (1)(c) does not apply in the case of a foreign national who
(a) has been determined to be a member of the family class and to be the spouse, common-law partner or child of a sponsor within the meaning of the regulations;
(b) has applied for permanent residence visa as a Convention refugee or a person in similar circumstances;
(c) is a protected person; or
(d) is, where prescribed by the regulations, the spouse, common-law partner, child or other family member of a foreign national referred to in any of paragraphs (a) to (c).

Notes

Introduction: The Personal and the Political

1 Abdelmalek Sayad, *La double absence* (Paris: Seuil, 1999), 396.
2 Michael Oliver, *Understanding Disability* (New York: St Martin's Press, 1996), 5.
3 Edward W. Said, *Covering Islam: How the Media and the Experts Determine How We See the Rest of the World* (New York: Pantheon Books, 1981), 154.
4 Linda Peake, "'Race' and Sexuality: Challenging the Patriarchal Structuring of Urban Social Space," *Environment and Planning D: Society and Space* 11 (1993): 419. See also Howard Zinn, *Failure to Quit: Reflections of an Optimistic Historian* (Cambridge, MA: South End Press, 1993), 30.
5 Titchkosky, *Reading & Writing Disability Differently: The Textured Life of Embodiment* (Toronto: University of Toronto Press, 2007), 32.
6 Zinn, *Failure to Quit*, 30. See also Teun A. van Dijk, "Principles of Critical Discourse Analysis," *Discourse & Society* 4, no. 2 (1993): 253–4.
7 Oliver, *Understanding Disability*, 13.
8 Oliver, 166.
9 Howard Adelman, "The New Immigration Regulations," posted 22 January 2002, https://yorkspace.library.yorku.ca/xmlui/bitstream/handle/10315/2595/H%20A%20The%20New%20Immigration%20Regulations.pdf?sequence=1&isAllowed=y, 1.
10 Moira Escott, Officer at the Consulate General of Canada, letter to the author, 10 May 2007.
11 Moira Escott, Officer at the Consulate General of Canada, letter to the author, 10 May 2007.
12 Howard Adelman, "The New Immigration Regulations," 9.
13 In this book, I use the concept of labour-power to indicate that particular commodity that, in the words of Rosa Luxemburg, "is brought to market by those who possess no means of production of their own with which to produce other commodities."

14 Beth A. Ferri and David J. Connor, *Reading Resistance* (New York: Peter Lang Publishing, 2006), 17.
15 I argue that the same holds true for Canadian scholars.
16 Douglas Baynton, "Disability and the Justification of Inequality in American History," in *The New Disability History,* eds. Paul K. Longmore and Lauri Umansky (New York: New York University Press, 2001), 52.
17 Paul Longmore, *Why I Burned My Book and Other Essays on Disability* (Philadelphia: Temple University Press, 2003) 54–5.
18 Dustin Galer, "A Friend in Need or a Business Indeed?: Disabled Bodies and Fraternalism in Victorian Ontario," *Labour/Le Travail* 66 (Fall 2010): 13.
19 Parin Dossa, *Racialized Bodies, Disabling Worlds* (Toronto: University of Toronto Press, 2009), 4.
20 I use the term "civil society" as referring to social structures other than and in between the household and the state – for instance, different interest groups, associations, and organizations.
21 Andrea Mayr, "Introduction: Power, Discourse and Institutions," *Language and Power,* ed. Andrea Mayr (London: Continuum International Publishing Group, 2008) 7, 8.
22 Toronto, Vancouver, and Montreal are the three urban centres that have traditionally received the bulk of immigration.
23 Larry Bourne and Damaris Rose, "The Changing Face of Canada: The Uneven Geographies of Population and Social Change," *The Canadian Geographer* 45, no.1 (2001): 111.

1 The *Right* Citizen

1 Thomas E. Yingling, *AIDS and the National Body* (Princeton, NJ: Princeton University Press, 1997), 25.
2 Dossa, *Racialized Bodies, Disabling Worlds,* 96, 153.
3 Catherine A. Holland, *The Body Politic* (New York and London: Routledge, 2001), 66–8.
4 Susan Wendell, *The Rejected Body* (New York and London: Routledge, 1996), 41.
5 Claudia Malacrida, "Income Support Policy in Canada and the UK: Different, but Much the Same," *Disability & Society* 2, no. 6 (2010), 684.
6 Titchkosky, *Reading & Writing,* 132.
7 Ena Chadha, "'Mentally Defective' Not Welcome: Mental Disability in Canadian Immigration Law, 1859–1927," *Disability Studies Quarterly* 28, no. 1 (2008).
8 Dossa, *Racialized Bodies, Disabling Worlds,* 22.
9 Bryan S. Turner, *The Body and Society* (London: Sage Publications, 1996), 161.
10 Michel Foucault, *Psychiatric Power: Lectures at the College de France, 1973–1974* (New York: Palgrave MacMillan, 2006), 14. See also Michel Foucault,

"Society Must Be Defended": Lectures at the College the France, 1975–1976 (New York: Picador, 2003), 30.

11 Amy Fairchild, *Science at the Borders* (Baltimore: The John Hopkins University Press, 2003), 146.

12 Reg Whitaker, *Canadian Immigration Policy since Confederation* (Ottawa: Canadian Historical Association, 1991), 4.

13 Mabel F. Timlin, "Canada's Immigration Policy, 1896–1910," *The Canadian Journal of Economics and Political Science* 26, no 4 (1960), 517.

14 Donald Avery, *Reluctant Host: Canada's Response to Immigrant Workers 1896–1994* (Toronto: McClelland and Stewart, 1995), 10.

15 Timlin, "Canada's Immigration Policy," 518.

16 Avery, *Reluctant Host,* 10.

17 Fairchild, *Science,* 145.

18 Fairchild, 145–6.

19 Robert Menzies, "Governing Mentalities: The Deportation of 'Insane' and 'Feebleminded' Immigrants Out of British Columbia from Confederation to World War II," *Canadian Journal of Law and Society* 13, no. 2 (1998): 141–2.

20 Government of Canada. "Medical Inadmissibility," last modified 21 December 2018, https://www.canada.ca/en/immigration-refugees-citizenship/services/immigrate-canada/inadmissibility/reasons/medical-inadmissibility.html.

21 Teun A. van Dijk, "Principles of Critical Discourse Analysis," *Discourse & Society* 4, no. 2 (1993): 255.

22 Neil Boyd, "Law and Social Control, Law as Social Control," in *The Social Dimensions of Law,* ed. Neil Boyd (Scarborough, ON: Prentice-Hall Canada, 1986), 18.

23 Michel Foucault, *"Society Must Be Defended,"* 27.

24 Susan Wendell, *The Rejected Body,* 21.

25 Wendell, 20.

26 Dossa, *Racialized Bodies, Disabling Worlds,* 5.

27 Bill Hughes and Kevin Paterson, "The Social Model of Disability and the Disappearing Body: Towards a Sociology of Impairment," *Disability & Society* 12, no. 3 (1997), 328. See also Claire Tregaskis, "Social Model Theory: The Story so Far...," *Disability & Society* 17, no. 4 (2002): 459; Dossa, *Racialized Bodies, Disabling Worlds,* 16–17, 152.

28 Titchkosky, *Reading & Writing,* 38.

29 Hughes and Paterson, "The Social Model of Disability," 329.

30 Liz Crow, "Including All of Our Lives: Renewing the Social Model of Disability," in *Exploring the Divide: Illness and Disability,* eds. Colin Barnes and Geof Mercer (Leeds: The Disability Press, 1996), 59.

31 Howard Zinn, *Howard Zinn Speaks: Collected Speeches 1963–2009,* ed. Anthony Arnove (Chicago: Haymarket Books, 2012), 94–8.

32 Said, *Covering Islam*, 154–7.
33 Zinn, *Failure to Quit*, 30.
34 For a discussion of history as relationship of force see Michel Foucault, *"Society Must Be Defended,"* 168–70.
35 Himani Bannerji, "Introducing Racism: Towards an Anti-Racist Feminism," in *Open Boundaries: A Canadian Women's Studies Reader*, eds. Barbara A. Crow and Lise Gotell (Toronto: Prentice-Hall Canada, 2000), 30.
36 Audra Jennings, "Engendering and Regendering Disability: Gender and Disability Activism in Postwar America," in *Disability Histories*, eds. Susan Burch and Michael Rembis (Urbana: University of Illinois, 2014), 346.
37 David M. Ryfe, "The Principles of Public Discourse: What Is Good Public Discourse?," in *Public Discourse in America*, eds. Judith Rodin and Stephen Steinberg (Philadelphia: University of Pennsylvania Press, 2003), 163.
38 Martin J. Burke, *The Conundrum of Class* (Chicago: The University of Chicago Press, 1995), xiii.
39 Thomas Bender, "The Thinning of American Political Culture," in *Public Discourse in America*, eds. Judith Rodin and Stephen Steinberg (Philadelphia: University of Pennsylvania Press, 2003), 33.
40 Neil Smelser, "A Paradox of Public Discourse and Political Democracy," in *Public Discourse in America*, eds. Judith Rodin and Stephen Steinberg (Philadelphia: University of Pennsylvania Press, 2003), 182.
41 van Dijk, "Principles of Critical Discourse Analysis," 255–7, 260.
42 Burke, *The Conundrum of Class*, xiii.
43 Burke, xv.
44 Burke, xvi.
45 Mayr, "Introduction: Power, Discourse and Institutions," 5.
46 Karim H. Karim, "Reconstructing the Multicultural Community in Canada: Discursive Strategies of Inclusion and Exclusion," *International Journal of Politics, Culture and Society* 7, no. 2 (1993): 193.
47 Colin Barnes, "Theories of Disability and the Origins of the Oppression of Disabled People in Western Society," in *Disability and Society*, ed. Len Barton (New York: Longman Publishing, 1996), 52–3.
48 John Swain, Sally French, and Colin Cameron, *Controversial Issues in a Disabling Society* (Philadelphia: Open University Press, 2003), 23.
49 Michael Foucault, *Madness and Civilization* (New York: Vintage Book, 1965), 46.
50 Foucault, 47.
51 Rosemarie G. Thomson, *Extraordinary Bodies* (New York: Columbia University Press, 1997), 39.
52 Foucault, *Psychiatric Power*, 112.
53 Foucault, *Madness and Civilization*, 56.
54 Peter Conrad and Joseph W. Schneider, *Deviance and Medicalization* (St. Louis: The C.V. Mosby Company, 1980), v.

55 Conrad and Schneider, 249.
56 Titchkosky, *Reading & Writing*, 38.
57 Titchkosky, 55–6, 105.
58 Dossa, *Racialized Bodies, Disabling Worlds*, 79.
59 Daniel T. Rodgers, *The Work Ethic in Industrial America, 1850–1920* (Chicago: The University of Chicago Press, 1974), 226.
60 Dustin Galer, "Packing Disability into the Historian's Toolbox: On the Merits of Labour Histories of Disability," *Labour/Le Travail* 77 (Spring 2016): 260.
61 Samuel Gridley Howe, as cited in James W. Trent Jr, *Inventing the Feeble Mind* (Los Angeles: University of California Press, 1994), 24.
62 Galer, "A Friend in Need," 11.
63 Brad Byrom, "A Pupil and a Patient: Hospital-Schools in Progressive America," in *The New Disability History*, eds. Paul K. Longmore and Lauri Umansky (New York: New York University Press, 2001), 135.
64 Jennings, "Engendering and Regendering Disability," 346.
65 James E. Moran, *Committed to the State Asylum* (Montreal: McGill-Queen's University Press, 2000), 49, 62.
66 Russell C. Smandych and Simon N. Verdun-Jones, "The Emergence of the Asylum in 19th Century Ontario: A Study in the History of Segregative Control," in *The Social Dimensions of Law*, ed. Neil Boyd (Scarborough, ON: Prentice-Hall Canada, 1986), 171.
67 Smandych and Verdun-Jones, 171.
68 Moran, *Committed*, 93.
69 Moran, 68.
70 Geoffrey Reaume, "Patients at Work: Insane Asylum Inmates' Labour in Ontario, 1841–1900," in *Mental Health and Canadian Society: Historical Perspectives*, eds James E. Moran and David Wright (Montreal: McGill-Queen's University Press, 2006), 71.
71 Reaume, 77, 83–4.
72 Reaume, 77.
73 James W. Trent, *Inventing the Feeble Mind* (Los Angeles: University of California Press, 1994) 275–7.
74 Colin Barnes and Geof Mercer, "Disability, Work, and Welfare: Challenging the Social Exclusion of Disabled People," *Work, Employment and Society* 19, no. 3 (2005): 533.
75 Barnes and Mercer, 541.
76 Oliver, *Understanding Disability*, 127.
77 Oliver, 76.
78 Rachel Hurst, "Conclusion: Enabling or Disabling Globalization?," in *Controversial Issues in a Disabling Society*, eds. John Swain, Sally French, and Colin Cameron (Buckingham, PA: Open University Press, 2003), 166.

79 Michael Oliver, "Disability and Dependency: A Creation of Industrial Societies," in *Disability and Dependency*, ed. Len Barton (London: The Falmer Press, 1989), 11.
80 Oliver, 11. See also Dossa, *Racialized Bodies, Disabling Worlds*, 13, and Titchkosky, *Reading & Writing*, 157.
81 Dossa, 153.
82 Barnes and Mercer, "Disability, Work, and Welfare," 532.
83 Vera Chouinard, "Body Politics: Disabled Women's Activism in Canada and Beyond," in *Mind and Body Spaces*, eds. Ruth Butler and Hester Parr (London: Routledge, 1999), 269, 282–5.
84 Council of Canadians with Disabilities, "Make EI Accessible and Inclusive to Canadian Women with Disabilities," posted 31 March 2009, http://www.ccdonline.ca/en/socialpolicy/employment/EI-pressrelease-31March2009.
85 Michael J. Prince, *Absent Citizens: Disability Politics and Policy in Canada* (Toronto: University of Toronto Press, 2009), 23, 209.
86 Prince, 4.
87 Lyn Jongbloed, "Disability Policy in Canada: An Overview," *Journal of Disability Policy Studies* 13 (2003): 208.
88 Chouinard, "Body Politics," 285, 289.
89 Malacrida, "Income Support Policy," 684.
90 Wendell, *The Rejected Body*, 13.
91 Robert Proctor, *Racial Hygiene: Medicine under the Nazis* (Cambridge, MA: Harvard University Press, 1988), 177–8.
92 Trent, *Inventing the Feeble Mind*, 134.
93 Proctor, *Racial Hygiene*, 185.
94 Jenny Morris, *Pride Against Prejudice* (London: The Women's Press, 1991), 132.
95 Proctor, *Racial Hygiene*, 180. See also Sunera Thobani, *Exalted Subjects* (Toronto: University of Toronto Press, 2007), 150.
96 Paul K. Longmore, "Elizabeth Bouvia, Assisted Suicide, and Social Prejudice," in *Why I Burned My Book and Other Essays on Disability* (Philadelphia: Temple University Press, 2003), 153.
97 Longmore, 153.
98 Margaret Lock, "Genomics, Laissez-Faire Eugenics, and Disability," in *Disability in Local and Global Worlds*, eds. Benedicte Ingstad and Susan Reynolds Whyte (Berkeley: University of California Press, 2007), 193, 205. See also Dossa, *Racialized Bodies, Disabling Worlds*, 42
99 Dossa, 77.
100 Morris, *Pride Against Prejudice*, 132.
101 Susan E. Brown, Debra Connors, and Nancy Stern, *With the Power of Each Breath* (Pittsburg: Cleis Press, 1985), 14.
102 Lock, "Genomics," 204.

103 Alan Sears, "Immigration Control as Social Policy: The Case of Canadian Medical Inspection 1900–1920," *Studies in Political Economy* 33 (Autumn 1990): 91.
104 Menzies "Governing Mentalities," 140.
105 Menzies, 140.
106 Whitaker *Canadian Immigration Policy,* 4.
107 Chadha, "'Mentally Defective' Not Welcome."
108 Ninette Kelley and Michael Trebilcock, *The Making of the Mosaic* (Toronto: University of Toronto Press, 1998), 13.
109 Martin Pâquet, *Tracer les marges de la cité: Étranger, immigrant et l'État au Québec, 1627–1981* (Montreal: Boreal, 2005), 152.
110 Patricia K. Wood, "Defining 'Canadian': Anti-Americanism and Identity in Sir John A. Macdonald's Nationalism," *Journal of Canadian Studies* 36, no. 2 (2001): 63.
111 Harold Troper, *History of Immigration since the Second World War* (Toronto: Joint Centre of Excellence for Research on Immigration and Settlement, 2000), 7.
112 Whitaker, *Canadian Immigration Policy,* 8.
113 Pâquet, *Tracer les marges,* 142–3.
114 Titchkosky, *Writing & Reading,* 149.
115 *An Act to Amend the Immigration Act,* R.S.C. 65, 2E VII, c. 14, s. 24.
116 Chadha, "'Mentally Defective' Not Welcome."
117 Whitaker, *Canadian Immigration Policy,* 11.
118 *Immigration Act, 1906,* https://pier21.ca/research/immigration-history/immigration-act-1906.
119 Chadha, "'Mentally Defective' Not Welcome."
120 James Shaver Woodsworth, *Strangers within Our Gates* (Toronto: University of Toronto Press, 1972), 223.
121 William George Smith, *Building the Nation: A Study of Some Problems Concerning the Churches' Relation to the Immigrants* (Toronto: Canadian Council of The Missionary Education Movement, 1922), 84–5.
122 Patricia T. Rooke and R.L. Schnell, *Discarding the Asylum* (Lanham, MD: University Press of America, 1983) 230.
123 Rooke and Schnell, 230.
124 Deborah Carter Park and John Radford, "Rhetoric and Place in the 'Mental Deficiency' Asylum," in *Mind and Body Spaces,* eds. Ruth Butler and Hester Parr (London: Routledge, 1999), 92.
125 Avery, *Reluctant Host,* 84–5.
126 Claudia Malacrida, "Discipline and Dehumanization in a Total Institution: Institutional Survivors' Description of Time-Out Rooms," *Disability & Society* 20, no. 5 (2005): 526.
127 Chadha, "'Mentally Defective' Not Welcome."

128 William Lyon Mackenzie King, as cited in Whitaker, *Canadian Immigration Policy*, 14.
129 Freda Hawkins, *Canada and Immigration* (Montreal: McGill-Queen's University Press, 1972), 101–2.
130 Chadha, "'Mentally Defective' Not Welcome."
131 Dossa, *Racialized Bodies, Disabling Worlds*, 3. See also Abu-Laban, "welcome/ STAY OUT: The Contradiction of Canadian Integration and Immigration Policies at the Millennium," in "Canadian Immigration," supplement, *Canadian Ethnic Studies* 30, no. 3 (1998): 192.
132 Whitaker, *Canadian Immigration Policy*, 19.
133 Troper, *History of Immigration*, 45. See also Whitaker, 19.

2 Parliament and Medically Inadmissible Immigrants

1 Immigration and Refugee Protection Act, art. 38(1)(c).
2 Angus McLaren, *Our Own Master Race* (Toronto: McClelland & Stewart, 1990), 25–7.
3 Patricia E. Roy, *The Triumph of Citizenship* (Vancouver: UBC Press, 2007), 14.
4 Roy, 308.
5 van Dijk, "Principles of Critical Discourse Analysis," 268.
6 Pâquet, *Tracer les marges*, 140.
7 *House of Commons Debates*, 9th Parliament, 2nd Session, 29 April 1902, 3743.
8 Troper, *History of Immigration*, 7.
9 *House of Commons Debates*, 9th Parliament, 2nd Session, 29 April 1902, 3743.
10 *House of Commons Debates*, 9th Parliament, 3rd Session, 14 July 1903, 6550.
11 *House of Commons Debates*, 9th Parliament, 3rd Session,14 July 1903, 6610.
12 *House of Commons Debates*, 9th Parliament, 4th Session, 21 July 1904, 7285.
13 *Senate Debates*, 10th Parliament, 1st Session, 8 June 1905, 367–8.
14 *Sessional Papers*, 1905, vol. X no. 25. Robin: Would sessional papers be the same?
15 *Sessional Papers*, 1905, vol. X no. 25.
16 *House of Commons Debates*, 11th Parliament, 1st Session, 10 May 1909, 6129.
17 *Senate Debates*, 10th Parliament, 1st Session, 8 June 1905, 368.
18 Annual Report of the Department of the Interior for the Fiscal Year Ended 30th June, 1904, *Sessional Paper No. 25*.
19 *House of Commons Debates*, 10th Parliament, 2nd Session, 13 June 1906, 5246.
20 Dossa, *Racialized Bodies, Disabling Worlds*, 32.
21 *House of Commons Debates*, 10th Parliament, 2nd Session, 13 June 1906, 5247.
22 *House of Commons Debates*, 10th Parliament, 2nd Session, 13 June 1906, 5247.
23 *Senate Debates*, 10th Parliament, 2nd Session, 3 July 1906, 1083.
24 *Senate Debates*, 10th Parliament, 2nd Session, 3 July 1906, 1079.

25 *Senate Debates*, 10th Parliament, 2nd Session, 3 July 1906, 1081.
26 *Senate Debates*, 10th Parliament, 2nd Session, 3 July 1906, 1080.
27 Ian R. Dowbiggin, *Keeping America Sane* (Ithaca, NY: Cornell University Press, 2003), 147.
28 *House of Commons Debates*, 10th Parliament, 2nd Session, 13 June 1906, 5248.
29 *House of Commons Debates*, 10th Parliament, 2nd Session, 13 June 1906, 5247.
30 *House of Commons Debates*, 10th Parliament, 2nd Session, 13 June 1906, 5250.
31 *House of Commons Debates*, 10th Parliament, 2nd Session, 13 June 1906, 5250.
32 *Senate Debates*, 10th Parliament, 2nd Session, 3 July 1906, 1085.
33 *House of Commons Debates*, 10th Parliament, 3rd Session, 3 April 1907, 5718.
34 *House of Commons Debates*, 10th Parliament, 3rd Session, 3 April 1907, 5719.
35 Menzies, "Governing Mentalities," 157.
36 *House of Commons Debates*, 11th Parliament, 2nd Session, 14 March 1910, 5519.
37 *House of Commons Debates*, 11th Parliament, 2nd Session, 14 March 1910, 5520.
38 *House of Commons Debates*, 11th Parliament, 2nd Session, 14 March 1910, 5520.
39 Annual Report of the Department of the Interior for the Fiscal Year 1908–9, *Sessional Paper No.25.*
40 McLaren, *Our Own Master Race*, 52.
41 Annual Report of the Department of the Interior for the Fiscal Year 1908–9, *Sessional Paper No. 25.*
42 Annual Report of the Department of the Interior for the Fiscal Year 1914, *Sessional Paper No. 25.*
43 Sunera Thobani, *Exalted Subjects* (Toronto: University of Toronto Press, 2007), 3–4.
44 McLaren, *Our Own Master Race*, 57.
45 Avery, *Reluctant Host*, 84–5. See also Malacrida, "Discipline and Dehumanization," 526.
46 McLaren, *Our Own Master Race*, 94.
47 *House of Commons Debates*, 13th Parliament, 2nd Session, 29 April 1919, 1872.
48 *House of Commons Debates*, 13th Parliament, 2nd Session, 7 April 1919, 1207
49 *House of Commons Debates*, 13th Parliament, 2nd Session, 30 April 1919, 1945.
50 *House of Commons Debates*, 13th Parliament, 2nd Session, 1 May 1919, 1974.
51 *House of Commons Debates*, 16th Parliament, 2nd Session, 1 March 1928, 909.
52 *House of Commons Debates*, 16th Parliament, 2nd Session, 1 March 1928, 909.
53 Thobani, *Exalted Subjects*, 76.
54 McLaren, *Our Own Master Race*, 66.
55 In 1912 the Moral and Social Reform Council, a national and provincial organization created in 1908 by Protestant churches adhering to the Social Gospel, changed its name to Social Service Council of Canada.

56 *House of Commons Debates*, 16th Parliament, 2nd Session, 6 June 1928, 3809.
57 Obligatory inspection overseas was only introduced in 1928. See McLaren, *Our Own Master Race*, 65.
58 *House of Commons Debates*, 16th Parliament, 2nd Session, 5 March 1928, 1001.
59 *House of Commons Debates*, 16th Parliament, 2nd Session, 9 March 1928, 1186.
60 *House of Commons Debates*, 16th Parliament, 2nd Session, 12 March 1928, 1229.
61 *House of Commons Debates*, 16th Parliament, 2nd Session, 12 March 1928, 1229.
62 *House of Commons Debates*, 16th Parliament, 2nd Session, 12 March 1928, 1204.
63 *House of Commons Debates*, 16th Parliament, 2nd Session, 12 March 1928, 1234.
64 *House of Commons Debates*, 16th Parliament, 2nd Session, 12 March 1928, 1234.
65 *House of Commons Debates*, 16th Parliament, 2nd Session, 12 March 1928, 1234.
66 *House of Commons Debates*, 16th Parliament, 2nd Session, 8 June 1928, 3949.
67 *House of Commons Debates*, 16th Parliament, 2nd Session, 8 June 1928, 3968.
68 *House of Commons Debates*, 16th Parliament, 2nd Session, 8 June 1928, 3968.
69 *House of Commons Debates*, 17th Parliament, 2nd Session, 10 July 1931, 3647.
70 Menzies "Governing Mentalities," 138.
71 Menzies, 171. Dowbiggin, *Keeping America Sane*, 178.
72 *Senate Debates*, 18th Parliament, 3rd Session, 4 April 1938, 212.
73 *Senate Debates*, 18th Parliament, 3rd Session, 4 April 1938, 211.
74 Roy, *The Triumph of Citizenship*, 10.
75 *House of Commons Debates*, 21st Parliament, 7th Session, 24 April 1953, 4356.
76 Menzies, "Governing Mentalities," 170. See also Thobani, *Exalted Subjects*, 150–1. It should be noted that this holds true for Canada overall, although there were exceptions, as in the case of Alberta where eugenics theories continued to be popular until the 1970s. See Jana Grekul, Harvey Krahn, and Dave Odynak, "Sterilizing the "Feeble-minded": Eugenics in Alberta, Canada, 1929–1972," *Journal of Historical Sociology* 17, no. 4 (2004): 358–84.
77 Alvin Finkel, *Social Policy and Practice in Canada: A History* (Waterloo: Wilfred Laurier University Press, 2006), 170.
78 Mary Tremblay, "Lieutenant John Counsell and the Development of Medical Rehabilitation and Disability Policy in Canada," in *Making Equality*, eds. Deborah Stienstra and Aileen Wight-Felske (Toronto: Captus Press, 2003), 59.
79 Tremblay, 65.
80 Jeff Keshen, "Getting It Right the Second Time Around," in *The Veteran Charter and Post–World War II Canada*, eds. Peter Neary and J.L. Granatstein (Montreal: McGill-Queen's University Press, 1998), 77.
81 Keshen, 76–8.
82 *House of Commons Debates*, 21st Parliament, 7th Session, 28 November 1952, 180.
83 *Senate Debates*, 22nd Parliament, 1st Session, 17 June 1954.
84 *House of Commons Debates*, 22nd Parliament, 3rd Session, 15 June 1956, 5095.

85 *House of Commons Debates*, 22nd Parliament, 3rd Session, 15 June 1956, 5095.

86 Alfred Neufeld, "Disability Policy (Canada)," *Encyclopedia of Social Welfare History in North America*, eds. John Herrick and Paul Stuart (Thousand Oaks: Sage Publications, 2005), 79.

87 *House of Commons Debates*, 29th Parliament, 1st Session, 19 February 1973, 1423.

88 *House of Commons Debates*, 29th Parliament, 1st Session, 19 February 1973, 1423.

89 The National Health Program was created in 1948, followed by the Unemployment Assistance Act in 1956 and the Hospital Insurance and Diagnostic Services Act in 1957.

90 Thobani, *Exalted Subjects*, 108; Dossa, *Racialized Bodies, Disabling Worlds*, 84.

91 Roy, *The Triumph of Citizenship*, 187–8.

92 Hawkins, *Canada and Immigration*, 101–2.

93 Avery, *Reluctant Host*, 176–8.

94 Raymond Plant, *The Neo-Liberal State* (Oxford: Oxford University Press, 2010), 117. See also David J. Roberts and Minelle Mahtani, "Neoliberalizing Race, Racing Neoliberalism: Placing 'Race' in Neoliberal Discourses," *Antipode* 42, no. 2 (2010): 253–5.

95 Roberts and Mahtani, 251–3.

96 Dossa, *Racialized Bodies, Disabling Worlds*, 22.

97 *House of Commons Debates*, 27th Parliament, 2nd Session, 22 March 1968, 7974–81.

98 *Senate Debates*, 27th Parliament, 2nd Session, 26 March 1968, 1013.

99 *House of Commons Debates*, 27th Parliament, 1st Session, 25 January 1966, 259.

100 *House of Commons Debates*, 27th Parliament, 2nd Session, 22 March 1968, 7977.

101 *House of Commons Debates*, 28th Parliament, 1st Session, 22 November 1968, 3114.

102 *House of Commons Debates*, 28th Parliament, 1st Session, 22 November 1968, 3116.

103 *Senate Debates*, 30th Parliament, 1st Session, 6 November 1975, 1385.

104 *Senate Debates*, 30th Parliament, 1st Session, 6 November 1975, 1385.

105 *Senate Debates*, 30th Parliament, 1st Session, 6 November 1975, 1386.

106 *Senate Debates*, 30th Parliament, 2nd Session, 1 August 1977, 1174.

107 *House of Commons Debates*, 30th Parliament, 2nd Session, 21 March 1977, 4182.

108 *House of Commons Debates*, 30th Parliament, 2nd Session, 14 March 1977, 3955.

109 *House of Commons Debates*, 30th Parliament, 2nd Session, 14 March 1977, 3955.

110 *House of Commons Debates,* 30th Parliament, 2nd Session, 14 March 1977, 3955.
111 *House of Commons Debates,* 30th Parliament, 2nd Session, 14 March 1977, 3955.
112 *House of Commons Debates,* 30th Parliament, 2nd Session, 14 March 1977, 3955.
113 The medical practitioner is today's equivalent of the 1977 medical officer.
114 *House of Commons Debates,* 33rd Parliament, 1st Session, 28 March 1985, 3453.
115 *House of Commons Debates,* 34th Parliament, 2nd Session, 8 June 1989, 2751, and 34th Parliament, 2nd Session, 15 December 1989, 7000.
116 Whitaker, *Canadian Immigration Policy,* 23. Christopher P. Manfredi, *Judicial Power and the Charter: Canada and the Paradox of Liberal Constitutionalism* (Don Mills, ON: Oxford University Press, 2001), 125.
117 *House of Commons Debates,* 34th Parliament, 3rd Session, 19 June 1992, 12466.
118 *House of Commons Debates,* 34th Parliament, 3rd Session, 8 June 1992, 11550.
119 *House of Commons Debates,* 34th Parliament, 3rd Session, 8 June 1992, 11550.
120 *House of Commons Debates,* 34th Parliament, 3rd Session, 8 June 1992, 11580.
121 *House of Commons Debates,* 34th Parliament, 3rd Session, 8 June 1992, 11584.
122 *House of Commons Debates,* 34th Parliament, 3rd Session, 19 June 1992, 12466.
123 *Senate Debates,* 34th Parliament, 3rd Session, 15 December 1992, 2447–8.
124 Bill C-86/Projet de loi C-86, 2:13.
125 Bill C-86/Projet de loi C-86, 2:13.
126 Bill C-86/Projet de loi C-86, 2:13.
127 Bill C-86/Projet de loi C-86, 3:46.
128 Bill C-86/Projet de loi C-86, 3:52.
129 Bill C-86/Projet de loi C-86, 3:53.
130 In 1994 the Coalition of Provincial Organizations of the Handicapped changed its name to Council of Canadians with Disabilities.
131 Bill C-86/Projet de loi C-86, 4:80.
132 Bill C-86/Projet de loi C-86, 4:81.
133 Bill C-86/Projet de loi C-86, 4:81–2.
134 Bill C-86/Projet de loi C-86, 4:82.
135 Bill C-86/Projet de loi C-86, 4:82.
136 Bill C-86/Projet de loi C-86, 4:82.
137 Bill C-86/Projet de loi C-86, 4:82.
138 Bill C-86/Projet de loi C-86, 4:82.
139 Bill C-86/Projet de loi C-86, 4:83.
140 Bill C-86/Projet de loi C-86, 4:84.
141 Bill C-86/Projet de loi C-86, 4:85–6.
142 Bill C-86/Projet de loi C-86, 4:88.
143 Bill C-86/ Projet de loi C-86, 4A:106.

144 Bill C-86/Projet de loi C-86, 4A:94.
145 Bill C-86/Projet de loi C-86, 4:88–9.
146 Bill C-86/Projet de loi C-86, 4:90.
147 Bill C-86/Projet de loi C-86, 4:90.
148 Bill C-86/Projet de loi C-86, 4A:79.
149 Bill C-86/Projet de loi C-86, 4A:79.
150 Bill C-86/Projet de loi C-86, 4:91.
151 James S. Woodsworth, *Strangers within Our Gates* (Toronto: University of Toronto Press, 1909).
152 Thobani, *Exalted Subjects*, 4.
153 Bill C-86/Projet de loi C-86, 6:6.
154 Bill C-86/Projet de loi C-86, 6:6.
155 Bill C-86/Projet de loi C-86, 6:17.
156 Bill C-86/Projet de loi C-86, 6:17.
157 Bill C-86/Projet de loi C-86, 6:17.
158 Bill C-86/Projet de loi C-86, 4A:100.
159 Bill C-86/Projet de loi C-86, 11:9.
160 Bill C-86/Projet de loi C-86, 11:10.
161 Bill C-86/Projet de loi C-86, 11:16.
162 Bill C-86/Projet de loi C-86, 11:17.
163 Bill C-86/Projet de loi C-86, 11:10.
164 Bill C-86/Projet de loi C-86, 11:10.
165 Bill C-86/Projet de loi C-86, 11:20.
166 Bill C-86/Projet de loi C-86, 11:20.
167 Bill C-86/Projet de loi C-86, 11:20.
168 Bill C-86/Projet de loi C-86, 11:20.
169 Bill C-86/Projet d loi C-86, 11:21.
170 Bill C-86/Projet d loi C-86, 11:21.
171 Bill C-86/Projet de loi C-86, 11:24.
172 Abu-Laban, welcome/STAY OUT," 194.
173 *House of Commons Debates*, 35th Parliament, 1st Session, 2 February 1994, 803.
174 *House of Commons Debates*, 35th Parliament, 1st Session, 7 June 1994, 4949.
175 *House of Commons Debates*, 35th Parliament, 1st Session, 7 June 1994, 4949.
176 Thobani, *Exalted Subjects*, 180.
177 Alan Simmons, "Immigration Policy: Imagined Futures," in *Immigrant Canada: Demographic, Economic and Social Challenges*, eds. Shiva Halli and Leo Driedger (Toronto: University of Toronto Press, 1999), 41.
178 House of Commons Debates, 15 April 1994, 3102.
179 *House of Commons Debates*, 35th Parliament, 1st Session, 15 April 1994, 3102.
180 *House of Commons Debates*, 35th Parliament, 1st Session, 23 September 1994, 6104.

181 *House of Commons Debates*, 35th Parliament, 1st Session, 23 September 1994, 6105.
182 *House of Commons Debates*, 35th Parliament, 1st Session, 23 September 1994, 6105.
183 *House of Commons Debates*, 35th Parliament, 1st Session, 23 September 1994, 6105.
184 *House of Commons Debates*, 35th Parliament, 1st Session, 23 September 1994, 6105.
185 *House of Commons Debates*, 35th Parliament, 1st Session, 23 September 1994, 6106.
186 *House of Commons Debates*, 35th Parliament, 1st Session, 23 September 1994, 6106.
187 *House of Commons Debates*, 35th Parliament, 1st Session, 23 September 1994, 6106.
188 *House of Commons Debates*, 35th Parliament, 1st Session, 23 September 1994, 6107.
189 *House of Commons Debates*, 35th Parliament, 1st Session, 23 September 1994, 6107.
190 The party rejected the motion.
191 *House of Commons Debates*, 35th Parliament, 1st Session, 24 October 1994, 7065.
192 *House of Commons Debates*, 35th Parliament, 1st Session, 24 October 1994, 7066.
193 *House of Commons Debates*, 35th Parliament, 1st Session, 24 October 1994, 7067.
194 *House of Commons Debates*, 35th Parliament, 1st Session, 24 October 1994, 7068.
195 *House of Commons Debates*, 35th Parliament, 1st Session, 24 October 1994, 7071.
196 *House of Commons Debates*, 35th Parliament, 1st Session, 24 October 1994, 7071.
197 *House of Commons Debates*, 35th Parliament, 1st Session, 24 October 1994, 7071.
198 *House of Commons Debates*, 35th Parliament, 1st Session, 31 October 1994, 7395.
199 *House of Commons Debates*, 35th Parliament, 1st Session, 31 October 1994, 7396.
200 *House of Commons Debates*, 35th Parliament, 1st Session, 31 October 1994, 7397.
201 *House of Commons Debates*, 35th Parliament, 1st Session, 20 February 1995, 9825.
202 *House of Commons Debates*, 35th Parliament, 1st Session, 20 February 1995, 9825.
203 *House of Commons Debates*, 35th Parliament, 1st Session, 20 February 1995, 9825.
204 Paul Hunt, "A Critical Condition," in *The Disability Reader*, ed. Tom Shakespeare (London: Continuum, 1998), 11.
205 Thobani, *Exalted Subjects*, 180–3.
206 Titchkosky, *Reading & Writing*, 141.
207 Standing Committee on Citizenship and Immigration, 8 June 2000, http://www.ourcommons.ca/DocumentViewer/en/36-2/CIMM/meeting-30/evidence.

208 Standing Committee on Citizenship and Immigration, 8 June 2000.
209 Standing Committee on Citizenship and Immigration, 8 June 2000.
210 Standing Committee on Citizenship and Immigration, 29 March 2000, http://www.ourcommons.ca/DocumentViewer/en/36-2/CIMM/meeting-19/evidence.
211 Standing Committee on Citizenship and Immigration, 29 March 2000.
212 Same title as the previous bill.
213 Bill C-11: The Immigration and Refugee Protection Act, http://www.parl.gc.ca/common/bills_ls.asp?lang=E&ls=c11&source=library_prb&Parl=37&Ses=1.
214 Standing Committee on Citizenship and Immigration, 1 May 2001. http://www.ourcommons.ca/DocumentViewer/en/37-1/CIMM/meeting-14/evidence.
215 Standing Committee on Citizenship and Immigration, 1 May 2001.
216 Standing Committee on Citizenship and Immigration, 1 May 2001.
217 Standing Committee on Citizenship and Immigration, 1 May 2001.
218 Standing Committee on Citizenship and Immigration, 1 May 2001.
219 Standing Committee on Citizenship and Immigration, 1 May 2001.
220 Standing Committee on Citizenship and Immigration, 1 May 2001.
221 Standing Committee on Citizenship and Immigration, 1 May 2001.
222 Standing Committee on Citizenship and Immigration, 1 May 2001.
223 Standing Committee on Citizenship and Immigration, 1 May 2001.
224 Standing Committee on Citizenship and Immigration, 5 February 2002. http://www.ourcommons.ca/DocumentViewer/en/37-1/CIMM/meeting-45/evidence.
225 Standing Committee on Citizenship and Immigration, 5 February 2002.
226 Standing Committee on Citizenship and Immigration, 5 February 2002.
227 Standing Committee on Citizenship and Immigration, 5 February 2002.
228 Standing Committee on Citizenship and Immigration, 5 February 2002.
229 Standing Committee on Citizenship and Immigration, 12 March 2002. http://www.ourcommons.ca/DocumentViewer/en/37-1/CIMM/meeting-53/evidence.
230 Standing Committee on Citizenship and Immigration, 12 March 2002.
231 Jacques Derrida, *On Cosmopolitanism and Forgiveness* (London: Routledge, 2001), 4.
232 Hawkins, *Canada and Immigration*, 347.
233 Roy, *The Triumph of Citizenship*, 308.
234 Hawkins, *Canada and Immigration*, 351.
235 Titchkosky, *Reading & Writing*, 47.
236 Didier Fassin, “Coming Back to Life: An Anthropological Reassessment of Biopolitics and Governmentality,” in *Governmentality*, eds. Ulrich Bröckling, Susanne Krasmann, and Thomas Lemke (New York: Routledge, 2011), 185.

237 Thomas Lemke, "Beyond Foucault: From Biopolitics to the Government of Life," in *Governmentality*, eds. Ulrich Bröckling, Susanne Krasmann and Thomas Lemke (New York: Routledge, 2011), 166. See also Michel Foucault, *"Society Must Be Defended,"* 244–5.
238 Lemke, "Beyond Foucault," 177–8.

3 Medical Admissibility: *Toronto Star* and the *Globe and Mail*, 1902–1985

1 Jay Rosen, "Part of Our World: Journalism as Civic Leadership," in *Public Discourse in America*, eds. Judith Rodin and Stephen P. Steinberg (Philadelphia: University of Pennsylvania Press, 2003), 108.
2 Rosen, 108.
3 David Hayes, *Power and Influence* (Toronto: Key Porter Books, 1992), 49.
4 Bourne and Rose, "The Changing Face of Canada," 110.
5 Avery, *Reluctant Host*, 13.
6 *Toronto Star*, 1 July 1967, 54.
7 Mohammad Qadeer and Sandeep Kumar, "Ethnic Enclaves and Social Cohesion," *Canadian Journal of Urban Research* 15, no. 2 (2006): 5.
8 Statistics Canada, *Canada's Ethnocultural Mosaic, 2006 Census* (Ottawa: Minister of Industry, 2008), 29.
9 Paul Rutherford, *The Making of the Canadian Media* (Toronto: McGraw-Hill Ryerson Limited, 1978), 49.
10 Effie Ginzberg, *Power without Responsibility: The Press We Don't Deserve* (Toronto: Urban Alliance on Race Relations, 1987), 7–8.
11 van Dijk, "Principles of Critical Discourse Analysis," 253.
12 Said, *Covering Islam*, 44.
13 Said, 45.
14 Robert Hackett, *News and Dissent: The Press and the Politics of Peace in Canada* (Norwood, NJ: Ablex Pub. Corp., 1991), 70–1. See also Said, *Covering Islam*, 49.
15 Hayes, *Power and Influence*, 7.
16 Des Freedman, *The Politics of Media Policy* (Cambridge: Polity Press, 2008), 6–7.
17 Ferri and Connor, *Reading Resistance*, 15.
18 Said *Covering Islam*, 45–6. See also Philo, "Can Discourse Analysis Successfully Explain," 181 and van Dijk, "Principles of Critical Discourse Analysis," 280.
19 Said, *Covering Islam*, 46.
20 van Dijk "Principles of Critical Discourse Analysis," 260.
21 Frances Henry, *Discourses of Domination: Racial Bias in the Canadian English-Language Press* (Toronto: University of Toronto Press, 2002), 7.
22 Hackett 53. See also Philo, "Can Discourse Analysis Successfully Explain," 181.

23 Philo, 182.
24 Hayes, *Power and Influence*, 6, 58–9, 69.
25 Said *Covering Islam*, 45.
26 "Excluding Undesirables," *Globe*, 21 January 1910, 4.
27 "The Immigration Act," *Globe*, 2 May 1902, 4.
28 *Globel*, 22 August 1902, 4.
29 Avery, *Reluctant Host*, 10–11.
30 "They Go through Canada," *Globe*, 13 December 1902, 12.
31 "Our Immigration Policy," *Globe*, 16 July 1903, 6.
32 "Get Good Value in Immigration," *Toronto Star*, 22 July 1904, 7.
33 "Roosevelt Seeks Our Cooperation," *Toronto Star*, 5 December 1905, 4.
34 "Stop Poisoning the National Blood," *Toronto Star*, 21 February 1905, 6.
35 "Stop Poisoning the National Blood," 6.
36 "Deporting the Insane," *Globe*, 30 October 1906, 6.
37 "Deporting the Insane," 6.
38 "Canada's Future Citizens," *Globe*, 17 July 1907, 6.
39 "Canada's Future Citizens," 6.
40 "Canada's Future Citizens," 6.
41 "No Undesirables Are Wanted Here," *Globe*, 15 November 1907, 12.
42 "Mr Hanna Slanders the Foreign-Born," *Globe*, 19 September 1908, 4.
43 "Mr Hanna Slanders the Foreign-Born," 4.
44 "Excluding Undesirables," *Globe*, 21 January 1910, 4.
45 "Excluding Undesirables," 4.
46 "Selecting immigrants," *Toronto Star*, 15 June 1909, 6.
47 "The Prevention of Pauperism," *Toronto Star*, 28 November 1910, 8.
48 "The Prevention of Pauperism," 8.
49 "Six Women to Be Deported from Toronto Institutions," *Globe*, 28 August 1913, 8.
50 "Six Women to Be Deported from Toronto Institutions," 8.
51 "Feeble-Minded a Ghastly Menace," *Globe*, 1 December 1916, 6.
52 "Feeble-Minded a Ghastly Menace," 6.
53 "Feeble-Minded a Ghastly Menace," 6.
54 "Plead Proper Inspections," *Globe*, 27 February 1918, 5.
55 In 1950 the Canadian National Committee for Mental Hygiene was renamed Canadian Mental Health Association.
56 "Guard against Weak-Minded," *Globe*, 7 December 1918, 10.
57 "City Council Favors Plan to Care for Feeble-Minded," *Toronto Star*, 19 December 1916, 5.
58 "Would Ban Defectives," *Toronto Star*, 29 September 1917, 1.
59 "Improve Immigration," *Toronto Star*, 16 January 1919, 5.
60 McLaren, *Our Own Master Race*, 64–5.
61 McLaren, 65; Dowbiggin, *Keeping America Sane*, 148–9, 151.

62 C.K. Clarke, "The Defective Immigrant," *Globe*, 27 January 1919, 4.
63 C.K. Clarke, "The Defective Immigrant," 4.
64 C.K. Clarke, "The Defective Immigrant," 4.
65 "Would Bar Out Feeble-Minded from Dominion," *Globe*, 5 March 1920, 8.
66 "Canada Needs Ellis Island Says Expert," *Globe*, 13 November 1920, 17.
67 "Medical Inspectors Busy," *Toronto Star*, 25 May 1920, 12.
68 "The Examination of Emigrants," *Toronto Star*, 31 July 1920, 6.
69 "Immigrants Today Must Pass Tests of Mind and Body," *Globe*, 9 March 1923, 12.
70 "Many New Immigrants Are Public Charges," *Globe*, 13 February 1924, 13.
71 "Many New Immigrants Are Public Charges," 13.
72 "Stricter Selection of All Immigrants Is Strongly Urged," *Globe*, 29 January 1925, 3.
73 "Feeble-Minded Come from Lands Outside of Canada," *Globe*, 26 August 1925, 9.
74 "Feeble-Minded Come from Lands Outside of Canada," 9.
75 The Dominion Education Association had changed its name to Canadian Education Association in 1918.
76 "Immigration Barrier Is Not Tight Enough," *Globe*, 12 November 1925, 2.
77 The National Council of Women was founded in Toronto in 1893 with the goal of joining together women of different creeds, churches, and races.
78 "Mental Defectives Are Not Canadians," *Toronto Star*, 15 June 1921, 22.
79 "Suggest Institutions for Defective Youths," *Toronto Star*, 16 February 1924, 16.
80 McLaren, *Our Own Master Race*, 167.
81 Avery, *Reluctant Host*, 84–5.
82 "Says Eugenics Greatly Affects Nation's Future," *Toronto Star*, 14 August 1924, 17.
83 "Says Eugenics Greatly Affects Nation's Future," 17.
84 "Urge Careful Choice of Child Immigrants," *Toronto Star*, 1 October 1924, 3.
85 "Urge Careful Choice of Child Immigrants," 3.
86 "Trustees Request Ottawa to Bar Mental Defectives," *Toronto Star*, 16 October 1925, 35.
87 "Immigration Needed, but Mental Fitness Must Be Demanded," *Globe*, 27 May 1926, 2.
88 "Immigration Needed, but Mental Fitness Must Be Demanded," 2.
89 Frank Chamberlain, "Immigration in Operation," *Globe*, 28 November 1927, 4.
90 Dowbiggin, *Keeping America Sane*, 134.
91 "Won't Rescind Order for Deportation," *Toronto Star*, 28 July 1926, 3.
92 "Betty Roy's Worries Over: Will Stay Here," *Toronto Star*, 4 August 1926, 2.
93 "Betty Roy's Worries Over: Will Stay Here," 2.
94 "Seek Right to Deport Naturalized Insane," *Toronto Star*, 16 February 1928, Z99.

95 Kanishka Jayasuriya, *Statecraft, Welfare and the Politics of Inclusion* (Houndmills: Palgrave Macmillan, 2006), 4.
96 Wendy Larner, "Neo-liberalism: Policy, Ideology, Governmentality," *Studies in Political Economy* 63 (Autumn 2000): 19.
97 Larner, 19.
98 "Bar Baby Immigrant from Joining Father," *Toronto Star*, 3 February 1928, 2.
99 "Bar Baby Immigrant from Joining Father," 2.
100 "Child Was Deficient – Deportation Reason," *Toronto Star*, 7 February 1928, 3.
101 "Ottawa Investigates Maconachie Baby Ban," *Toronto Star*, 11 February 1928, 11.
102 "Agitation Continues for Entry of Child," *Toronto Star*, 6 March 1928, 21.
103 "Agitation Continues for Entry of Child," 21.
104 "Child Barred by Law – McConachie Finding," *Toronto Star*, 10 March 1928, 22.
105 "A Case Calling for Sympathy," *Globe*, 27 February 1928, 4.
106 "Legally Impossible to Make Exception of McConachie Case," *Globe*, 29 February 1928, 5.
107 "Legally Impossible to Make Exception of McConachie Case," 5.
108 "McConachie Infant Declared Imbecile by Scottish Expert," *Globe*, 14 March 1928, 7.
109 "McConachie Infant Declared Imbecile by Scottish Expert," 7.
110 Foucault, *"Society Must Be Defended,"* 107.
111 "Says Scots Girl Nearly as Broad as She Was Tall," *Toronto Star*, 12 May 1928, 1.
112 "Says Scots Girl Nearly as Broad as She was Tall," 1.
113 *Globe*, 12 May 1928, 4.
114 "Why Closer Sifting is Needed," *Globe*, 14 May 1928, 4.
115 "Why Closer Sifting is Needed," 4.
116 "Medical Association Meets at Kingston," *Toronto Star*, 31 May 1928, 17.
117 "Say Entry Was Illegal," *Toronto Star*, 17 October 1930, 3.
118 "Say Entry Was Illegal," 3.
119 "Say Entry Was Illegal," 3.
120 "Hamilton Epileptic Missed Steamer: Deportation Likely to Be Appealed," *Toronto Star*, 25 October 1930, 1.
121 *Toronto Star*, 24 October 1930, 2.
122 "Hamilton Mother Asks Aid to Save Daughter from Being Deported," *Toronto Star*, 27 October 1930, 1.
123 "Parents Complain No Notice Was Given in Deportation Case," *Globe*, 30 October 1930, 2.
124 "Parents Complain No Notice Was Given in Deportation Case," 2.
125 "Parents Complain No Notice Was Given in Deportation Case," 2.
126 "Doctor's Statement under Investigation," *Globe*, 31 December 1930, 2.
127 "Deported to Old Land Woman Was Landed without Even Carfare," *Toronto Star*, 27 October 1930, 27.

128 “Deported to Old Land Woman Was Landed without Even Carfare,” 27.
129 “Immigration Official Denies Any Inhumanity in Deporting Unfit,” *Toronto Star*, 28 October 1930, 1.
130 *Toronto Star*, 28 October 1930, 2.
131 *Toronto Star*, 28 October 1930, 2.
132 “Deportation Monstrous is Bennett’s Assertion: Should Be Another Way,” *Toronto Star*, 31 October 1930, 1.
133 “Deportation Monstrous is Bennett’s Assertion,” 1.
134 “Deportation Monstrous is Bennett’s Assertion ,” 1.
135 “Deportation Monstrous is Bennett’s Assertion ,” 1.
136 “Deportation Monstrous is Bennett’s Assertion ,” 1.
137 “Deportation Monstrous is Bennett’s Assertion,” 1.
138 “Mother Deported and 4 Children Are Separated,” 1.
139 “Mother Deported and 4 Children Are Separated,” 1.
140 “Mother Deported and 4 Children Are Separated,” 1.
141 “Mother Deported and 4 Children Are Separated,” 1.
142 “Many Undesirables Admitted to Canada,” *Toronto Star*, 4 November 1930, 36.
143 “More Humanitarian Action Pledged in Mental Cases,” *Toronto Star*, 5 November 1930, 3.
144 “More Humanitarian Action Pledged in Mental Cases ,” 3.
145 “More Humanitarian Action Pledged in Mental Cases,”3.
146 Thobani, *Exalted Subjects*, 20–1.
147 “More Humanitarian Action Pledged in Mental Cases,” 3.
148 “Robb Passing Buck Hamilton Mayor Says,” *Toronto Star*, 5 November 1930, 1.
149 “Immigration Minister Should Quit: Gordon,” *Toronto Star*, 7 November 1930, 1.
150 “Deported Britishers ‘Dumped Like Cattle’ Bolton Alderman Says,” *Toronto Star*, 11 November 1930, 23.
151 “Immigration and Deportation,” *Globe*, 4 March 1931, 4.
152 “Immigration and Deportation,” 4.
153 “Immigration and Deportation,” 4.
154 Here a further distinction should be made between white and non-white imperial subjects.
155 “An Amazing Muddle,” *Globe*, 10 April 1931, 4
156 “Amend This Inhuman Law,” *Globe*, 23 April 1931, 4.
157 “Amend This Inhuman Law,” 4.
158 “Equal Rights Wanted,” *Globe*, 17 August 1931, 4.
159 Avery *Reluctant Host*, 13. See also Yves Engler, *Canada and Israel: Building Apartheid* (Vancouver: RED Publishing, 2010), 27.
160 “Canada Immigration Laws Too Lax, Professor Says,” *Toronto Star*, 1 June 1931, 20.
161 “Canada Immigration Laws Too Lax,” *Toronto Star*, 1 June 1931, 20.

162 “Canada Immigration Laws Too Lax,” 20.
163 “Canada Immigration Laws Too Lax,” 20.
164 “Canada Immigration Laws Too Lax,” 20.
165 “Immigration Men Harsh with Girl Mother Alleges,” *Toronto Star*, 29 December 1930, 1.
166 “Girl’s Deportation Ordered: Ottawa Gives Two Reasons,” *Toronto Star*, 31 December 1930, 30.
167 “Canada’s Red Tape Impedes Romance,” *Toronto Star*, 18 February 1932, 23.
168 “Stars Said to Marry, Deafness Bars Bride,” *Toronto Star*, 23 April 1934, 7.
169 Newfoundland Would Only Join Confederation in 1949.
170 “Man’s Home, Job at Stake: Bar Child’s Canada Entry,” *Toronto Star*, 15 February 1946, 17.
171 “Man’s Home, Job at Stake,” 17.
172 “Man’s Home, Job at Stake,” 17.
173 “Man’s Home, Job at Stake,” 17.
174 “MD’s Reprieve keeps Harjit Here – For Now,” *Toronto Star*, 27 January 1980, A1.
175 Avery *Reluctant Host*, 145.
176 “Humans – or Superhumans?,” *Globe and Mail*, 6 March 1952, 6.
177 “Canada Rejects 12th Child – Silver Wedding Delayed,” *Toronto Star*, 11 May 1954, 12.
178 “Can’t Enter Canada,” *Toronto Star*, 19 December 1956, 6.
179 “Wife Excluded,” *Toronto Star*, 8 May 1957, 6.
180 “Wife Excluded,” 6.
181 “Wife Excluded,” 6.
182 “Bar Irish Family of 6: One Child Is Epileptic,” *Toronto Star*, 23 October 1958, 31.
183 “Bar Irish Family of 6,” 31.
184 “Arthritic Denied Entry to Canada,” *Toronto Star*, 28 March 1959, 3.
185 “Remove This Threat,” *Toronto Star*, 2 March 1959, 6.
186 “Refugees: Are We Smug and Selfish?” *Toronto Star*, 17 December 1959, 6.
187 “Refugees: Are We Smug and Selfish?” 6.
188 *Toronto Star*, 19 December 1959, 1.
189 *Toronto Star*, 19 December 1959, 1.
190 *Toronto Star*, 19 December 1959, 1.
191 The Act went into effect on June 1, 1953.
192 Hawkins, *Canada and Immigration*, 101–2.
193 “Be Humanitarian: Admit Handicapped Immigrant,” *Toronto Star*, 11 September 1964, 3.
194 “Be Humanitarian,” 3.
195 “Entry to Canada and U.S. Refused Mentally Defective,” *Globe and Mail*, 22 July 1965, W02.

196 The Canadian Association for Retarded Children was a federation of provincial associations formed in 1956.
197 "Entry to Canada and U.S. Refused Mentally Defective," W02.
198 "Entry to Canada and U.S. Refused Mentally Defective," W02.
199 Barry Smart, *Economy, Culture and Society* (Buckingham, PA: Open University Press, 2003), 74. See also van Dijk, "Principles of Critical Discourse Analysis," 263–4.
200 "Entry to Canada and U.S. Refused Mentally Defective," W02.
201 "Volunteer Agency Sought to Aid Mentally Retarded," *Globe and Mail*, 26 July 1965, 12.
202 "Volunteer Agency Sought to Aid Mentally Retarded," 12.
203 "Volunteer Agency Sought to Aid Mentally Retarded," 12.
204 "Plan Policy Change to Admit Retarded," *Globe and Mail*, 1 October 1965, 8.
205 The Canadian Psychiatric Association had been founded in 1951.
206 "Amend Immigration Laws on Mentally Ill, Psychiatrists Urge," *Globe and Mail*, 13 January 1966, 23.
207 "Amend Immigration Laws on Mentally Ill," 23.
208 "Mental Disorder Should Not Be Ground for Deportation: CMHA," *Globe and Mail*, 17 February 1967, 10.
209 "Mental Disorder Should Not Be Ground for Deportation," 10.
210 "Mental Disorder Should Not Be Ground for Deportation," 10.
211 "Mental Illness Is Not a Crime," *Toronto Star*, 1 March 1967, 6.
212 Avery *Reluctant Host*, 178–80.
213 "Compassion Beats the Rule Book," *Toronto Star*, 7 December 1968, 6.
214 "Compassion Beats the Rule Book," 6.
215 "Compassion Beats the Rule Book," 6.
216 "Fear of Deportation Haunts Her, Polio Girl Pleads for Citizenship," *Toronto Star*, 12 August 1972, 1.
217 "Polio Victim's University Hopes Jeopardized By Immigration Law," *Globe and Mail*, 14 August 1972, 5.
218 Hughes and Paterson, "The Social Model of Disability," 328.
219 "Girl Crippled by Polio to Get Landed Status," *Globe and Mail*, 17 August 1972, W04; "Polio Victim Shouts for Joy as Mackasey Lets Her Stay," *Toronto Star*, 17 August 1972, 3.
220 "Blind Girl Ordered Deported Despite Pleas to Mackasey," *Toronto Star*, 6 November 1972, 1.
221 "Lynn Gets a Job – and Curt Letter Kicking Her Out," *Toronto Star*, 6 November 1972, 25.
222 "Mackasey Lets Blind Girl Stay Here," *Toronto Star*, 8 November 1972, 1.
223 "Mackasey Won't Expel Blind Woman," *Globe and Mail*, 8 November 1972, 1.
224 "Mackasey Lets Blind Girl Stay Here," 1.
225 Thobani, *Exalted Subjects*, 20–2.

226 "Taiwan Doctor Faces Deportation on Friday," *Globe and Mail*, 20 June 1973, 13.
227 "CMHA Protests against Decision to Deport Doctor," *Globe and Mail*, 21 June 1973, 45.
228 "Doctor Given 30 Days to Appeal Deportation," *Globe and Mail*, 22 June 1973, 13.
229 "Taiwan Doctor wins Immigration Appeal," *Globe and Mail*, 8 October 1974, 14.
230 "Alaskan Retarded Son Refused Entry Despite Job Offer," *Globe and Mail*, 10 December 1974, 13.
231 The National Institute for Mental Retardation was launched in 1970 by the Canadian Association for Community Living.
232 "Ottawa Allows Retarded Child into Canada," *Globe and Mail*, 13 December 1974, 13.
233 "Alaskan Retarded Son Refused Entry Despite Job Offer," 13.
234 "Further Wait for Retarded Boy in Alaska," *Globe and Mail*, 14 December 1974, 15.
235 Smart, *Economy, Culture and Society*, 83.
236 Jayasuriya, *Statecraft, Welfare and the Politics of Inclusion*, 5.
237 "MP Fights Sick Man's Deportation," *Toronto Star*, 8 March 1975, A3.
238 "Andras to Review Case of Man Deported for Mental Illness," *Toronto Star*, 11 March 1975, A8.
239 "Special Andras Permit Allows Fijian Ordered Deported to Stay," *Toronto Star*, 14 March 1975, A4.
240 "Health Rules with an Unhealthy Effect," *Toronto Star*, 11 September 1975, B4.
241 "Mother Still Fighting Deportation Order," *Globe and Mail*, 21 October 1975, 12.
242 "Group Protests Decision to Deport Retarded Girl," *Globe and Mail*, 6 December 1975, 4.
243 "Let Retarded Child Return to Canada," *Toronto Star*, 8 December 1975, C4.
244 "Inflexible," *Globe and Mail*, 9 December 1975, 6.
245 "Immigration Fight: The Winner Says No," *Globe and Mail*, 22 January 1976, 1.
246 "Immigration Fight," 1.
247 "Immigration Fight," 1.
248 "Immigrant Status Rejected by Epileptic," *Globe and Mail*, 22 January 1976, 2.
249 "Waiting for Pluck," *Globe and Mail*, 23 January 1976, 6.
250 "Changes Sought in Medical Bars to Immigrants," *Globe and Mail*, 27 January 1976, 43.
251 "Changes Sought in Medical Bars to Immigrants," 43.
252 "End Discrimination against Epileptics," *Toronto Star*, 2 February 1976, C4.

253 Jacob S. Ziegel and Donald C. Savage, "Epilepsy," letter to the editor, *Globe and Mail*, 1 March 1976, 6.
254 "Early Reform of Immigration Rules on Epileptics Unlikely," *Globe and Mail*, 13 February 1976, 5.
255 "Epileptic Rejects Second Offer, Wants Immigration Law Changed," *Globe and Mail*, 12 March 1976, 4.
256 "Updating Immigration Law," *Globe and Mail*, 30 November 1976, 6.
257 "Updating Immigration Law," 6.
258 "Hong Kong Family Split by Little Boy's Plight," *Globe and Mail*, 8 August 1977, 10.
259 The act was passed on 5 August 1977.
260 Thobani, *Exalted Subjects*, 96–8.
261 "Paraplegic Fights Deportation Order," *Toronto Star*, 29 November 1978, B14.
262 "Departure Order to Be Ignored, Paraplegic from Ecuador," *Globe and Mail*, 11 December 1978, 9.
263 "Departure Order to Be Ignored," 9.
264 "Departure Order to Be Ignored," 9.
265 "Blind Woman to Be Deported," *Toronto Star*, 8 December 1979, A5.
266 "Blind Woman Given Time to Find Care," *Toronto Star*, 12 December 1979, p. A12; "Blind Woman Gets Extension on Visiting Visa," *Globe and Mail*, 12 December 1979, 9.
267 Ron Santana, "Blind Woman's Plight," letter to the editor, *Globe and Mail*, 28 December 1979, 7.
268 Veerendra Adhiya, letter to the editor, *Toronto Star*, 2 January 1980, A9.
269 Veerendra Adhiya, A9.
270 Ferri and Connor, *Reading Resistance*, 147.
271 van Dijk "Principles of Critical Discourse Analysis," 256.
272 Avery, *Reluctant Host*, 171.
273 "Atkey Promises to Review Blind Woman's Deportation," *Toronto Star*, 15 January 1980, A2.
274 "No Appeal for Illegal Immigrants," *Toronto Star*, 17 January 1980, A9.
275 "Atkey Gives Sick Woman a Last-Minute Reprieve," *Toronto Star*, 25 January 1980, A3.
276 "Atkey Gives Sick Woman a Last-Minute Reprieve," A3.
277 "MD's Reprieve Keeps Harji Here – for Now," *Toronto Star*, 27 January 1980, A1.
278 "Immigration Antics Lack Heart," *Toronto Star*, 30 January 1980, A8.
279 Ron Atkey, letter to the editor, *Toronto Star*, 7 February 1980, A9.
280 "She Married to Stay in Canada but Harjit Still Faces Expulsion," *Toronto Star*, 6 February 1980, A1.
281 Peter Rickwood, "Indian Bride Turns to U.S. for Help," *Toronto Star*, 10 February 1980, A2. The reference is to the US Embassy hostage crisis in Iran that began on 4 November 1979 and ended on 20 January 1981.

282 Barbara Keddy, "Threatened by Federal Officials in Expulsion Case, Doctor Says," *Globe and Mail*, 11 February 1980, 5.
283 "Atkey Orders Harjit Back to India," *Toronto Star*, 12 February 1980, A3.
284 Barbara Keddy, "Indian Woman Loses Her Fight against Ouster," *Globe and Mail*, 12 February 1980, 2.
285 "Atkey Orders Harjit Back to India," A3.
286 John Picton, "Harjit's Expulsion Has Cost Us $80,000," *Toronto Star*, 17 February 1980, A1.
287 A. Davidson, "Disgusted with All the Sympathy," letter to the editor, *Toronto Star*, 27 February 1980, A9.
288 Robert Sutton, "Jury Urges Ban," *Toronto Star*, 2 April 1980, C6.
289 Sutton, "Jury Urges Ban," C6.
290 Stephen Strauss, "Bar Schizophrenic Immigrants: Jury," *Globe and Mail*, 2 April 1980, 3.
291 Stephen Strauss, "Doctor Assails Ideas of Jury," *Globe and Mail*, 3 April 1980, 5.
292 Strauss, "Doctor Assails Ideas of Jury," 5.
293 Frank E. Cashman and Joel Jeffries, "Immigrant's Suicide," letter to the editor, *Globe and Mail*, 22 April 1980, 6.
294 Cashman and Jeffries, "Immigrant's Suicide," 6.
295 Dorothy Lipovenko, "Canada Shuts Door on Dying Man," *Globe and Mail*, 16 January 1981, 15.
296 Stephen Strauss, "Once Barred, Ill Man Arrives in Canada," *Globe and Mail*, 28 January 1981, 14.
297 Strauss, "Once Barred, Ill Man Arrives in Canada," 14.
298 Quebec joined in 1972.
299 Finkel, *Social Policy and Practice in Canada*, 186–9.
300 Richard Cleroux, "Manitoba Puts Forward Plan to Settle Handicapped Refugees," *Globe and Mail*, 7 July 1981, 11.
301 Richard Cleroux, "Manitoba Puts Forward Plan to Settle Handicapped Refugees," 11.
302 "Manitoba Proposal on Disabled Refugees Is Before Provinces," *Globe and Mail*, 8 July 1981, 2.
303 "Manitoba Proposal on Disabled Refugees Is Before Provinces," 2.
304 The reader should note that refugees are not considered immigrants under the Immigration Act; consequently, the refugee situation would require a separate book.
305 "Immigration Turns down Woman Because She's Deaf," *Toronto Star*, 24 February 1982, A3.
306 "Immigration Turns down Woman Because She's Deaf," A3.
307 Joe Serge, "Happy Ending for Flower Lady and Deaf Woman Seeking Entry," *Toronto Star*, 27 February 1982, A6

308 "Deaf Woman's Appeal Has Merit," *Toronto Star*, 1 March 1982, A14.
309 Sandro Contenta, "Heart Victim Can't Stay," *Toronto Star*, 27 February 1982, A1.
310 "Guyanese Man Can Stay for Operation," *Toronto Star*, 2 March 1982, A7.
311 Joe Serge, "'I Wouldn't Be a Burden to Canada' Says Polio Victim Who Wants to Stay," *Toronto Star*, 31 August 1983, C13.
312 Serge, "'I Wouldn't Be a Burden to Canada' Says Polio Victim Who Wants to Stay," C13.
313 Serge, "'I Wouldn't Be a Burden to Canada' Says Polio Victim Who Wants to Stay," C13.
314 Aldred H. Neufeldt, "Growth and Evolution of Disability Advocacy in Canada," in *Making Equality*, eds. Deborah Stienstra and Aileen Wight-Felske (Toronto: Captus Press, 2003), 19.
315 Finkel, *Social Policy and Practice in Canada*, 189.
316 Yvonne Peters, "From Charity to Equality," in *Making Equality*, eds. Deborah Stienstra and Aileen Wight-Felske (Toronto: Captus Press, 2003), 123.
317 John Myles and Feng Hou, "Changing Colours: Spatial Assimilation and New Racial Minority Immigrants," *Canadian Journal of Sociology* 29, no. 1 (2004): 30.
318 R. Alan Walks and Larry S. Bourne, "Ghettos in Canadian cities? Racial Segregation, Ethnic Enclaves and Poverty Concentration in Canadian Urban Areas," *The Canadian Geographer* 50, no. 3 (2006): 277.
319 Dossa, *Racialized Bodies, Disabling Worlds.*
320 Sherene Razack, *Looking White People in the Eye* (Toronto: University of Toronto Press, 1998), 136.
321 Dossa, *Racialized Bodies, Disabling Worlds*, 32.
322 Dossa, 32.

4 Medical Admissibility: *Toronto Star* and the *Globe and Mail*, 1985–2002

1 Jeff Sallot, "Tories to Alter Laws That Violate Charter," *Globe and Mail*, 1 February 1985, 8.
2 Freedman, *The Politics of Media* Policy, 1. See also Philo, "Can Discourse Analysis Successfully Explain," 181–2.
3 Sallot, "Tories to Alter Laws That Violate Charter," 1.
4 Sallot, "Tories to Alter Laws That Violate Charter," 8.
5 Sallot, "Tories to Alter Laws That Violate Charter," 8.
6 Sallot, "Tories to Alter Laws That Violate Charter," 8.
7 Dale Gibson, *The Law of the Charter: Equality Rights* (Toronto: Carswell, 1990), 45.

8 In 1987 the association changed its name to Canadian Association of Community Living.
9 Dorothy Lipovenko, “Immigration Law to Be Challenged,” *Globe and Mail*, 15 August 1985, M02.
10 Dorothy Lipovenko, “Immigration Law to Be Challenged,” M02.
11 As I will discuss in the next chapter, the question will be partially answered in 1989.
12 Joe Serge, “Pakistani Man ‘Shocked’ at False Report to Immigration Officials That He Had TB,” *Toronto Star*, 4 October 1985, A23.
13 Serge, “Pakistani Man ‘Shocked’ at False Report,” A23.
14 Serge, “Pakistani Man ‘Shocked’ at False Report,” A23.
15 Serge, “Pakistani Man ‘Shocked’ at False Report,” A23.
16 Robin Harvey, “Dieppe Veteran Barred from Canada Can Stay, Immigration Minister Says,” *Toronto Star*, 22 November 1985, A2.
17 The Royal Canadian Legion was created in 1925 to advocate on behalf of veterans and returned service members.
18 Victor Malarek, “Ailing Veteran Can’t Remain, Ottawa Decides,” *Globe and Mail*, 21 November 1985, A2.
19 Malarek, “Ailing Veteran Can’t Remain,” A2.
20 Malarek, “Ailing Veteran Can’t Remain,” A2.
21 Malarek, “Public Outcry Prompts Ottawa to Let Veteran Stay in Canada,” *Globe and Mail*, 22 November 1985, 2.
22 Malarek, “Public Outcry Prompts Ottawa to Let Veteran Stay in Canada,” 2.
23 Jayasuriya, *Statecraft, Welfare and the Politics of Inclusion*, 153.
24 “U.S. to Make AIDS Test a Must for Immigrants,” *Toronto Star*, 4 February 1986, A10.
25 “U.S. to Make AIDS Test a Must for Immigrants,” A10.
26 “U.S. to Make AIDS Test a Must for Immigrants,” A10.
27 Yingling, *AIDS and the National Body*, 25.
28 “Canada to Take Immigrants Exposed to AIDS Virus,” *Toronto Star*, 19 March 1986, A3.
29 “Canada to Take Immigrants Exposed to AIDS Virus,” A3.
30 Marina Strauss, “AIDS Test Proposal Is Called Premature,” *Globe and Mail*, 30 April 1986, A13.
31 Marina Strauss, “AIDS Test Proposal Is Called Premature,” A13.
32 Rose Voyvodic, “Into the Wasteland: Applying Equality Principles to Medical Inadmissibility in Canadian Immigration Law,” *Journal of Law and Social Policy* 16 (2001): 141.
33 Hannah Arendt, *The Origins of Totalitarianism* (New York: Harcourt Brace Jovanovich, 1973), 295.
34 Thobani, *Exalted Subjects*, 21.

35 "U.S. Orders AIDS Tests for All Immigrants," *Toronto Star*, 9 June 1987, A1.
36 "U.S. Orders AIDS Tests for All Immigrants," A1.
37 Victor Malarek, "3 Arrivals in Canada Suffering from AIDS, Health Official Reports," *Globe and Mail*, 10 June 1987, A13.
38 Ross Howard, "Panel Urges Mandatory AIDS Testing of Immigrants," *Globe and Mail*, 4 September 1987, A2.
39 Howard, "Panel Urges Mandatory AIDS Testing of Immigrants," A2.
40 Howard, "Panel Urges Mandatory AIDS Testing of Immigrants," A2.
41 Howard, "Panel Urges Mandatory AIDS Testing of Immigrants," A2.
42 Howard, "Panel Urges Mandatory AIDS Testing of Immigrants," A2.
43 Joan Breckenridge, "Testing of Immigrants Assailed as Ineffective," *Globe and Mail*, 10 November 1987, A3.
44 Breckenridge, "Testing of Immigrants Assailed as Ineffective," A3.
45 The government established the National Advisory Committee on AIDS in 1983 to advise the Minister of National Health and Welfare on how to control AIDS in Canada.
46 Breckenridge, "Testing of Immigrants Assailed as Ineffective," A3.
47 Breckenridge, "Testing of Immigrants Assailed as Ineffective," A3.
48 Breckenridge, "Testing of Immigrants Assailed as Ineffective," A3.
49 Christie McLaren, "Screening Immigrants Called a Way to Save Millions in Health Costs," *Globe and Mail*, 6 June 1989, A8.
50 McLaren, "Screening Immigrants Called a Way to Save," A8.
51 McLaren, "Screening Immigrants Called a Way to Save," A8.
52 McLaren, "Screening Immigrants Called a Way to Save," A8.
53 McLaren, "Screening Immigrants Called a Way to Save," A8.
54 Mike Funston, "Couple Fighting Losing Battle," *Toronto Star*, 3 March 1987, WE3.
55 Funston, "Couple Fighting Losing Battle," WE3.
56 Avery *Reluctant Host*, 222–3. See also Abu-Laban, "welcome/STAY OUT," 193–6.
57 "Immigration Association of Canada tells Parliament That Three Out of Four Canadians Want Immigration Placed on Hold until Federal Policies Are Clearly Stated and Approved by Majority of Canadians," *Globe and Mail*, 17 September 1987, A10.
58 "Immigration Association of Canada," A10.
59 "Immigration Association of Canada," A10.
60 "Immigration Association of Canada," A10.
61 Victor Malarek, "Ads on Immigration Intended to Spark Debate, Activist Says," *Globe and Mail*, 18 September 1987, A15.
62 Malarek, "Ads on Immigration Intended to Spark Debate," A15.
63 "Immigration Association of Canada," A10.

64 "Immigration Association of Canada," A10.
65 Sean Fine, "Man's Detention by U.S. Immigration Prompts Complaint," *Globe and Mail*, 22 October 1987, A18.
66 Fine, "Man's Detention by U.S. Immigration Prompts Complaint," A18.
67 Fine, "Man's Detention by U.S. Immigration Prompts Complaint,".
68 David Suzuki, "Scientists Should Not Turn from Twisted Nazi Genetics," *Globe and Mail*, 9 January 1988, D4.
69 David Suzuki, "Scientists Should Not Turn from Twisted Nazi Genetics," D4.
70 Suzuki, "Scientists Should Not Turn from Twisted Nazi Genetics," D4. For an analysis of Canada's strategy to distance itself from eugenics by adopting a multicultural policy in the post-war period, see Thobani, *Exalted Subjects*, 150–2.
71 David Suzuki, "Scientists Should Not Turn from Twisted Nazi Genetics," D4.
72 Suzuki, "Scientists Should Not Turn from Twisted Nazi Genetics," D4.
73 Ulrich Bröckling, "Human Economy, Human Capital: A Critique of Biopolitical Economy," in *Governmentality. Current Issues and Future Challenges*, eds., Ulrich Bröckling, Susanne Krasmann and Thomas Lemke (New York: Routledge, 2011), 259.
74 Crow, "Including All of Our Lives," 62.
75 Victor Malarek, "Immigration Department Fined over Mistreatment of Disabled Applicant," *Globe and Mail*, 6 March 1989, A1.
76 "Immigration Appealing Cash Award," *Toronto Star*, 8 March 1989, A8.
77 "Immigration Appealing Cash Award," A8.
78 Kazik Jedrzejczak, "Disabled Immigrant Treated Shamefully," letter to the editor, *Toronto Star*, 16 March 1989, A24.
79 "Case for Compassion," *Toronto Star*, 18 April 1989, A16.
80 Mary Bennett, "Giant Step Backward for Our Handicapped," letter to the editor, *Toronto Star*, 1 May 1989, A14.
81 Bonnie McDowell, "Bureaucratic Burden," letter to the editor, *Toronto Star*, 21 May 1989, B2.
82 Alfred Holden, "Peruvian Parents of Handicapped Boy Test Immigration Law," *Toronto Star*, 10 May 1989, A6.
83 Holden, "Peruvian Parents of Handicapped Boy Test Immigration Law," A6.
84 Alfred Holden, "Down Syndrome Shuts Out Foreign Families," *Toronto Star*, 19 May 1989, C1.
85 "Let's Fix This Law," *Toronto Star*, 22 May 1989, A14.
86 See previous chapter.
87 "Bar Irish Family of 6, One Child is Epileptic," *Toronto Star*, 23 October 1958, 31.
88 Thobani, *Exalted Subjects*, 20–1.
89 "A Gentle Apartheid," *Toronto Star*, 9 June 1989, A24.

90 "A Gentle Apartheid," A24.
91 Susan Reid, "Family of Down Girl Can Stay in Canada," *Toronto Star*, 20 July 1989, A3.
92 Julia Nunes, "MP Says Ottawa Likely to Alter Act to Remove Bias against Disabled," *Globe and Mail*, 21 July 1989, A3.
93 "A View from Edmonton: Fairness for Disabled," *Toronto Star*, 6 August 1989, B2.
94 "A View from Edmonton: Fairness for Disabled," B2.
95 Paul Watson, "Outdated Act Bars Immigrants Lawyers Say," *Toronto Star*, 17 February 1990, A3.
96 Watson, "Outdated Act Bars Immigrants Lawyers Say," A3.
97 Estanislao Oziewicz, "'Natural-Born' Canadian Denied Renewed Citizenship," *Globe and Mail*, 10 May 1991, A8.
98 Oziewicz, "'Natural-Born' Canadian Denied Renewed Citizenship," A8.
99 Oziewicz, "'Natural-Born' Canadian Denied Renewed Citizenship," A8.
100 Abu-Laban welcome/STAY OUT: 196–204; Thobani, *Exalted Subjects*, 180–3.
101 Avery, *Reluctant Host*, 230.
102 Hodgkin's disease is a cancer of the lymphatic system and is the third most common cancer among children in between ten and fourteen years of age.
103 Estanislao Oziewicz, "Family Faces Expulsion Because Son Has Cancer," A9.
104 Oziewicz, "Family Faces Expulsion Because Son Has Cancer," A9.
105 G.D. Elkin, "Lunacy Reign Supreme," letter to the editor, *Globe and Mail*, 9 October 1991, A16.
106 Elkin, "Lunacy Reign Supreme," A16.
107 Dan Irving, "Trans Politics and Anti-Capitalism: An Interview with Dan Irving," *Upping the Anti* 4 (May 2007): 70.
108 Alfons Mueller, "Invest in Human Life," letter to the editor, *Globe and Mail*, 10 October 1991, A20.
109 Mueller, "Invest in Human Life," A20.
110 Sharon Edmundson, "Simple Solution," letter to the editor, *Globe and Mail*, 6 November 1991, A17.
111 Edmundson, "Simple Solution," A17.
112 Estanislao Oziewicz, "Norwegian Family Allowed to Stay in Canada," *Globe and Mail*, 6 December 1991, A6.
113 Oziewicz, "Norwegian Family Allowed to Stay in Canada," A6.
114 Titchkosky, *Reading & Writing*, 167.
115 Barnes and Mercer, "Disability, Work, and Welfare," 541.
116 Paul Watson, "Immigration Rules May Limit Family Visas," *Toronto Star*, 6 November 1991, E14.
117 Watson, "Immigration Rules May Limit Family Visas," E14.
118 Watson, "Immigration Rules May Limit Family Visas," E14.
119 Watson, "Immigration Rules May Limit Family Visas," E14.

120 Rod Mickleburgh, "Stop Syphilis Tests for Immigrants, Panel Urges," *Globe and Mail*, 3 November 1992, A8.
121 Mickleburgh, "Stop Syphilis Tests for Immigrants," A8.
122 Mickleburgh, "Stop Syphilis Tests for Immigrants," A8.
123 Mickleburgh, "Stop Syphilis Tests for Immigrants," A8.
124 Mickleburgh, "Stop Syphilis Tests for Immigrants," A8.
125 Mickleburgh, "Stop Syphilis Tests for Immigrants," A8.
126 Mickleburgh, "Stop Syphilis Tests for Immigrants," A8.
127 Mickleburgh, "Stop Syphilis Tests for Immigrants," A8.
128 Allan Thompson, "HIV Tests Eyed for Immigrants," *Toronto Star*, 26 April 1994, A1.
129 Thompson, "HIV Tests Eyed for Immigrants," A1.
130 Thompson, "HIV Tests Eyed for Immigrants," A1.
131 "MP Seeks to Bar Immigrants with HIV," *Globe and Mail*, 24 September 1994, A3.
132 Lila Sarick, "Marchi Defends Refuge for HIV-Positive Pole," *Globe and Mail*, 21 February 1995, A4.
133 Sarick, "Marchi Defends Refuge for HIV-Positive Pole," A4.
134 Sarick, "Marchi Defends Refuge for HIV-Positive Pole," A4.
135 Sarick, "Marchi Defends Refuge for HIV-Positive Pole," A4.
136 Lisa Cherry, "Disabled Refugee Denied Citizenship," *Globe and Mail*, 2 September 1994, A4.
137 Cherry, "Disabled Refugee Denied Citizenship," A4.
138 According to Statistics Canada, the cost for an average Canadian was $12,370 over five years.
139 Cherry, "Disabled Refugee Denied Citizenship," A4.
140 Cherry, "Disabled Refugee Denied Citizenship," A4.
141 For a discussion on Bill C-86 see chapter 2.
142 Cherry, "Disabled Refugee Denied Citizenship," A4.
143 Cherry, "Disabled Refugee Denied Citizenship," A4.
144 Voyvodic, "Into the Wasteland," 29.
145 Allan Thompson, "Kids' Disabilities No Reason to Ban Immigrants: Judge," *Toronto Star*, 6 February 1996, A2.
146 Lila Sarick, "Recent Decisions Buoy Immigrants," *Globe and Mail*, 14 May 1996, A8.
147 Sarick, "Recent Decisions Buoy Immigrants," A8.
148 Sarick, "Recent Decisions Buoy Immigrants," A8.
149 Andre Picard, "Report Critical of Way Disabled Children Treated," *Globe and Mail*, 18 November 1999, A7.
150 Picard, "Report Critical of Way Disabled Children Treated," A7.
151 Allan Thompson, "No Entry for Immigrants with HIV," *Toronto Star*, 21 September 2000, A6.

152 The AIDS-HIV Legal Network was established in 1992 to promote human rights of persons with HIV/AIDS in Canada and internationally.
153 Thompson, "No Entry for Immigrants with HIV," A6.
154 Campbell Clark, "Immigrants facing blood tests," *Globe and Mail*, 21 September 2000, A4.
155 Clark, "Immigrants facing blood tests," A4.
156 Thompson, "No Entry for Immigrants with HIV," A6.
157 Thompson, "No Entry for Immigrants with HIV," A6.
158 Krista Yetman, "Costs would be too great," letter to the editor, *Toronto Star*, 26 September 2000, A27.
159 "HIV and immigration," *Globe and Mail*, 2 April 2001, A12.
160 Allan Thompson, "Ban on Immigrants with HIV Reversed," *Toronto Star*, 12 June 2001, A14.
161 Thompson, "Ban on Immigrants with HIV Reversed," A14.
162 Thompson, "Ban on Immigrants with HIV Reversed," A14.
163 Thompson, "Ban on Immigrants with HIV Reversed," A14.
164 Paul Taylor, "Wrong to Reverse HIV Ban," letter to the editor, *Toronto Star*, 13 June 2001, A31.
165 Taylor, "Wrong to Reverse HIV Ban," A31.
166 ARCH is a legal aid clinic for persons with disabilities. It was established in 1979 and operates in Ontario.
167 Maureen Murray, "Charging the Barricades," *Toronto Star*, 8 January 2002, A2.
168 Murray, "Charging the Barricades," A2.
169 Murray, "Charging the Barricades," A2.
170 Murray, "Charging the Barricades," A2.
171 Murray, "Charging the Barricades," A2.
172 Murray, "Charging the Barricades," A2.
173 Jane Gadd, "Immigration health rules challenged under Charter," *Globe and Mail*, 8 January 2002, A16.
174 Gadd, "Canada Contends It Tried to Help Disabled Woman," A12.
175 Gadd, "Canada Contends It Tried to Help Disabled Woman," A12.
176 Gadd, "Canada Contends It Tried to Help Disabled Woman," A12.
177 Gadd, "Canada Contends It Tried to Help Disabled Woman," A12.
178 Gadd, "Canada Contends It Tried to Help Disabled Woman," A12.
179 Gadd, "Canada Contends It Tried to Help Disabled Woman," A12.
180 *Toronto Star*, 10 January 2002, A20.
181 *Toronto Star*, 10 January 2002, A20.
182 Martha Rose, *The Staff of Oedipus* (Ann Arbor: The University of Michigan Press, 2003), 1.
183 Lynn Krzywiecki, "Let's show our humanity," letter to the editor, *Globe and Mail*, 9 January 2002, A10.
184 "Degrading Treatment," *Toronto Star*, 10 January 2002, A22.

185 "Welcome Change," *Toronto Star*, 1 July 2002, A14.
186 Razack, *Looking White People in the Eye*, 132.
187 Dossa, *Racialized Bodies, Disabling Worlds*, 22.
188 The issue of refugees is not entirely an internal policy matter since Canada has assumed binding responsibility vis-à-vis the international community.
189 Dossa, *Racialized Bodies, Disabling Worlds*, 23. See also Larner, "Neo-liberalism: Policy, Ideology, Governmentality," 19.
190 Gadd, "Canada contends it tried to help disabled woman," *Globe and Mail*, 9 January 2002, A12.
191 Gadd, "Canada contends it tried to help disabled woman," *Globe and Mail*, 9 January 2002, A12.

5 Medical Admissibility in the Federal and Supreme Courts of Canada

1 Sarah Armstrong, "Disability Advocacy in the Charter Era," *University of Toronto Journal of Law and Equality* 33 (2003).
2 Evelyn Kallen, *Ethnicity and Human Rights in Canada* (Toronto: Oxford University Press, 1995), 279. See also Gibson, *The Law of the Charter*, 147.
3 Kallen, *Ethnicity and Human Rights in Canada*, 261.
4 The Federal Court of Canada has two levels, the Federal Court Trial Division and the Federal Court of Appeal. Applications for judicial review of immigration decisions are heard by the Federal Court Trial Division; the judgment can then be appealed to the Federal Court of Appeal and then to the Supreme Court.
5 Pamar v. Canada (Minister of Employment and Immigration), No. A-836-87, Federal Court of Appeal, 16 May 1988.
6 *Pamar v. Canada.*
7 Deol v. Canada (Minister of Employment and Immigration), No. A-280-90, Federal Court of Appeal, 27 November 1992.
8 *Deol v. Canada.*
9 The Canadian Charter of Rights and Freedoms, Constitution Act, 1982.
10 *Deol v. Canada.*
11 Voyvodic, "Into the Wasteland," 142–3.
12 Giorgio Agamben, *State of Exception* (Chicago: The University of Chicago Press, 2005), 2–5.
13 Giorgio Agamben, *Homo Sacer: Sovereign Power and Bare Life* (Stanford, CA: Stanford University Press, 1995), 15–17.
14 Gingiovenanu v. Canada (Minister of Employment and Immigration), No. IMM-3875-93, Federal Court Trial Division, 30 October 1995.
15 Ismaili v. Canada (Minister of Citizenship and Immigration), No. IMM-3430-94, Federal Court Trial Division, 17 August 1995.
16 *Ismaili v. Canada.*

17 Poste v. Canada (Minister of Citizenship and Immigration), No. IMM-4601-96, Federal Court Trial Division, 22 December 1997.
18 *Poste v. Canada.*
19 *Poste v. Canada.*
20 *Poste v. Canada.*
21 Lau v. Canada (Minister of Citizenship and Immigration), No. IMM-4361-96, Federal Court Trial Division, 17 April 1998.
22 *Lau v. Canada.*
23 *Lau v. Canada.*
24 Poon v. Canada (Minister of Citizenship and Immigration), No. IMM-2007-99, Federal Court Trial Division, 1 December 2000.
25 *Poon v. Canada.*
26 Redding v. Canada (Minister of Citizenship and Immigration), No. IMM-2661-00, Federal Court Trial Division, 22 August 2001.
27 Wong v. Canada (Minister of Citizenship and Immigration), No. IMM-6060-99, Federal Court Trial Division, 31 May 2002.
28 *Wong v. Canada.*
29 Jayasuriya, *Statecraft, Welfare and the Politics of Inclusion*, 16.
30 Larner, "Neo-liberalism: Policy, Ideology, Governmentality," 19.
31 Hilewitz v. Canada (Minister of Citizenship and Immigration), No. IMM-5340-00, Federal Court Trial Division, 8 August 2002.
32 Hilewitz v. Canada (Minister of Citizenship and Immigration), No. A-560-02, Federal Court of Appeal, 12 November 2003.
33 *Hilewitz v. Canada*, 2003.
34 *Hilewitz v. Canada*, 2003.
35 *Hilewitz v. Canada*, 2003.
36 *Hilewitz v. Canada*, 2003.
37 Under the Immigration and Refugee Protection Act, ministerial permits have been renamed temporary residence permits.
38 Armstrong, "Disability Advocacy in the Charter Era."
39 Allan C. Hutchinson, *Waiting for Coraf: A Critique of Law and Rights* (Toronto: University of Toronto Press, 1995), 22.
40 Hutchinson, 24.
41 Hutchinson, 153.
42 Manfredi, *Judicial Power and the Charter*, 199.
43 Hutchinson, 56, 121.
44 Hutchinson, 224.
45 Hutchinson, 206.
46 Dossa, *Racialized Bodies, Disabling Worlds*, 66.
47 De Jong v. Canada (Minister of Citizenship and Immigration), Docket IMM-6058-99, Federal Court Trial Division, 13 November 2002.
48 *Deol v. Canada.*

49 Hilewitz v. Canada (Minister of Citizenship and Immigration); De Jong v. Canada (Minister of Citizenship and Immigration), No. 30125; 30127, Supreme Court of Canada, 21 October 2005.
50 *Hilewitz v. Canada* (2005); *De Jong v. Canada* (2005).
51 *Hilewitz v. Canada* (2005); *De Jong v. Canada* (2005).
52 *Hilewitz v. Canada* (2005); *De Jong v. Canada* (2005).
53 *Hilewitz v. Canada* (2005); *De Jong v. Canada* (2005).
54 *Hilewitz v. Canada* (2005); *De Jong v. Canada* (2005).
55 *Hilewitz v. Canada* (2005); *De Jong v. Canada* (2005).
56 *Hilewitz v. Canada* (2005); *De Jong v. Canada* (2005).
57 *Hilewitz v. Canada* (2005); *De Jong v. Canada* (2005).
58 *Hilewitz v. Canada* (2005); *De Jong v. Canada* (2005).
59 *Hilewitz v. Canada* (2005); *De Jong v. Canada* (2005).
60 *Hilewitz v. Canada* (2005); *De Jong v. Canada* (2005).
61 Immigration and Refugee Protection Act.
62 Ireh Iyioha, "A Different Picture through the Looking-Glass: Equality, Liberalism and the Question of Fairness in Canadian Immigration Health Policy," *Georgetown Immigration Law Journal* 22, no. 4 (2008): 632.
63 Iyioha, 632.
64 Larner, "Neo-liberalism: Policy, Ideology, Governmentality," 19.
65 Guidy Mamann, "CIC Skirts Court Medical Ruling," *Metro*, 10 September 2007.
66 Canada Constitution Act 1982, http://laws-lois.justice.gc.ca/eng/const/.
67 Operational Bulletin 037 – September 7. 2007. Assessing excessive demand on social services for business class applicants, http://www.cic.gc.ca/ENGLISH/resources/manuals/bulletins/2007/ob037.asp.
68 Guidy Mamann, "CIC Skirts Court Medical Ruling," *Metro*, 10 September 2007.
69 Mamann, Sandaluk & Kingwell LLP Immigration Layers http://www.migrationlaw.com/.
70 Armstrong, "Disability Advocacy in the Charter Era."
71 Armstrong.
72 Kallen, *Ethnicity and Human Rights in Canada*, 261.
73 Kallen, 261.
74 Chesters v. Canada (Minister of Citizenship and Immigration), No. IMM-1316-97, Federal Court Trial Division, 27 June 2002.
75 *Chesters v. Canada.*
76 *Chesters v. Canada.*
77 *Chesters v. Canada.*
78 Marika Willms, "Canadian Immigration Law and Same Sex Partners," *Canadian Issues/Thèmes Canadiens* (Spring 2005): 17.
79 The prothonotary is the chief clerk of the court in certain courts of law; the office exists in countries with Anglo-American jurisdiction.

80 Chesters v. Canada, Written Submissions of the Intervener, No. IMM-1316-97, Federal Court Trial Division, 27 February 2001.
81 *Chesters v. Canada,* Written Submissions of the Intervener.
82 *Chesters v. Canada,* Written Submissions of the Intervener.
83 *Chesters v. Canada,* https://www.canlii.org/en/ca/fct/doc/2002/2002fct727/2002fct727.html?autocompleteStr=Chester%20&autocompletePos=4.
84 *Chesters v. Canada.*
85 *Chesters v. Canada.*
86 Titchkosky, *Reading & Writing,* 58.
87 *Chesters v. Canada.*
88 *Chesters v. Canada.*
89 *Chesters v. Canada.*
90 *Chesters v. Canada.*
91 *Chesters v. Canada.*
92 Gibson, *The Law of the Charter,* 79.
93 Gibson, 79.
94 Voyvodic, "Into the Wasteland, 118, Manfredi, *Judicial Power and the Charter,* 124.
95 *Chesters v. Canada.*
96 *Chesters v. Canada.*
97 *Chesters v. Canada.*
98 *Chesters v. Canada.*
99 Armstrong, "Disability Advocacy in the Charter Era."
100 Voyvodic, "Into the Wasteland," 115, 128, 130.
101 Catherine Dauvergne, "Sovereignty, Migration and the Rule of Law in Global Times," *Modern Law Review* 67, no. 4 (July 2004): 591.
102 Armstrong, "Disability Advocacy in the Charter Era."
103 Dossa, *Racialized Bodies, Disabling Worlds,* 66.
104 Armstrong, "Disability Advocacy in the Charter Era."
105 William Boyce et al., *A Seat at the Table: Persons with Disabilities and Policy Making* (Montreal: McGill-Queen's University Press, 2001), 53.
106 Armstrong, "Disability Advocacy in the Charter Era."
107 In 1994, the Coalition of Provincial Organizations of the Handicapped changed its name to Council of Canadians with Disabilities.
108 Armstrong, "Disability Advocacy in the Charter Era."
109 Iyioha, "A Different Picture through the Looking-Glass," 659–60.
110 Peters, "From Charity to Equality 131.
111 Peters, 132.
112 Armstrong, "Disability Advocacy in the Charter Era."
113 Armstrong.
114 Armstrong.

115 Armstrong.
116 Arendt, *The Origins of Totalitarianism*, 293.
117 Giorgio Agamben, *Means without Ends* (Minneapolis: University of Minnesota Press, 2000), 19,0.
118 Catherine Frazee, “Exile from the China Shop: Cultural Injunction and Disability Policy,” in *Disability and Social Policy in Canada*, eds. Mary Ann McColl and Lyn Jongbloed (Concord, ON: Captus Press, 2006), 357. See also Iyioha, “A Different Picture through the Looking-Glass,” 625.
119 The document is enforceable only within the Canadian sovereign space.

Conclusion

1 Razack, *Looking White People in the Eye*, 136.
2 Dauvergne “Sovereignty, Migration and the Rule of Law in Global Times,” 590.
3 Kelley and Trebilcock, *The Making of the Mosaic*, 13.
4 Sears, “Immigration Control as Social Policy,” 91.
5 Wendy Brown, “We Are All Democrats Now...,” in *Democracy In What State?*, eds. Giorgio Agamben et al. (New York: Columbia University Press, 2011), 47.
6 Brown, 47.
7 Immigration and Refugee Protection Act, art. 3(1)(d) and (i).
8 Smart, *Economy, Culture and Society*, 141.
9 Mary Grimley Mason, *Working against Odds* (Boston: Northeastern University Press, 2004), 115, 118.
10 David Suzuki, *The David Suzuki Reader* (Vancouver: Greystone Books, 2003), 153.
11 Anderson, “The Idea of Chinatown: The Power of Place and Institutional Practice in the Making of a Racial Category,” in *Immigration in Canada: Historical Perspective*, eds. Gerald Tulchinsky (Toronto: Copp Clark Longman, 1994), 236.
12 Iris Marion Young, *Justice and the Politics of Difference* (Princeton, NJ: Princeton University Press, 1990), 227.
13 Titchkosky, *Reading & Writing*, 210.
14 Shearer, Disability: Whose Handicap? (Oxford: Basil Blackwell Publisher, 1981), 193.
15 Hunt, “A Critical Condition,” 13. Robin: The styling may have changed a bit here during editing (notes 15 and 16).
16 Sharon Dale Stone, “Resisting an Illness Label: Disability, Impairment, and Illness,” in *Contesting Illness. Processes and Practices*, eds. Pamela Moss and Katherine Teghtsoonian (Toronto: University of Toronto Press, 2008), 202.
17 United Nations: Enable. “Standard Rules on the Equalization of Opportunities for Persons with Disabilities,” http://www.un.org/esa/socdev/enable/dissre01.htm#Background.

18 Stone, "Resisting an Illness Label," 204–5.
19 Luna Bengio, *L'Article 19(1)A de la Loi Canadienne de l'Immigration* (Quebec: Association multi-ethnique pour l'intégration des personnes handicapées du Québec, 1990), 11.
20 In order to be accepted as sponsor, the individual must be either a Canadian citizen or permanent resident.
21 Agamben, *State of Exception*, 2–4.
22 Agamben, *State of Exception*, 87.
23 Agamben, *State of Exception*, 13, 22.
24 Agamben, *State of Exception*, 29–30.
25 Foucault, *Society Must Be Defended*, 245.
26 Foucault, *Society Must Be Defended*, 244
27 Foucault, *Society Must Be Defended*, 244.
28 Foucault, *Society Must Be Defended*, 254.
29 Agamben, *Homo Sacer*, 181.
30 Adelman, "The New Immigration Regulations," 9.
31 Peters, "From Charity to Equality," 134.
32 Daniel Bensaïd, "Permanent Scandal," in *Democracy in What State?*, eds. Giorgio Agamben et al. (New York: Columbia University Press, 2011), 41.
33 Valentina Capurri, "Women as Individuals and Members of Minority Groups: How to Reconcile Human Rights and the Values of Cultural Pluralism," *GeoJournal* 65, no. 4 (2006): 335.
34 Young, *Justice and the Politics of Difference*, 227.
35 Capurri, "Women as Individuals and Members of Minority Groups," 335.
36 Iyioha, "A Different Picture through the Looking-Glass," 650.
37 Judith K. Bernhard, Patricia Landolt, and Luin Goldring, "Transnationalizing Families: Canadian Immigration Policy and the Spatial Fragmentation of Care-giving among Latin American Newcomers," *International Migration* 47, no. 2 (2009): 6.
38 Arendt, *The Origins of Totalitarianism*, 296.
39 Michael J. Price, "Claiming a Disability Benefit as Contesting Social Citizenship," in *Contesting Illness: Processes and Practices*, eds. Pamela Moss and Katherine Teghtsoonian (Toronto: University of Toronto Press, 2008), 33.
40 Ferri and Connor, *Reading Resistance*, 121.
41 Engin Isin and Patricia Wood, *Citizenship and Identity* (London: Sage Publications, 1999), 22.
42 Judith Sandys, "Immigration and Settlement Issues for Ethno-Racial People with Disabilities," (1998), updated 1 September 1998, https://www.researchgate.net/publication/267549276_Immigration_And_Settlement_Issues_For_Ethno-Racial_People_With_Disabilities_An_Exploratory_Study.

Bibliography

Abberley, Paul. "Disabled People, Normality and Social Work." In *Disability and Dependency*, edited by Len Barton, 53–65. London: The Falmer Press, 1989.

Abu-Laban, Yasmeen. "welcome/STAY OUT: The Contradiction of Canadian Integration and Immigration Policies at the Millennium." *Canadian Ethnic Studies* 30, no. 3 (1998): 190–211.

Adelman, Howard. "The New Immigration Regulations." Posted 22 January 2002. https://yorkspace.library.yorku.ca/xmlui/bitstream/handle/10315/2595/H%20A%20The%20New%20Immigration%20Regulations.pdf?sequence=1&isAllowed=y.

Agamben, Giorgio. *Means without End: Notes on Politics.* Minneapolis: University of Minnesota Press, 2000.

– *State of Exception.* Chicago: The University of Chicago Press, 2005.

Anderson, Kay J. "The Idea of Chinatown: The Power of Place and Institutional Practice in the Making of a Racial Category." In *Immigration in Canada: Historical Perspective*, edited by Gerald Tulchinsky, 223–48. Toronto: Copp Clark Longman, 1994.

Arendt, Hannah. *The Origins of Totalitarianism*, New York: Harcourt Brace Jovanovich, 1973. First published 1951 by Schocken Books (New York City).

Avery, Donald. *Reluctant Host: Canada's Response to Immigrant Workers 1896–1994*, Toronto: McClelland and Stewart, 1995.

Bannerji, Himani. "Introducing Racism: Notes towards an Anti-Racism Feminism." In *Open Boundaries: A Canadian's Women Studies Reader*, edited by Barbara A. Crow and Lise Gotell, 26–32. Toronto: Prentice-Hall Canada, 2000.

Barnes, Colin. "Theories of Disability and the Origins of the Oppression of Disabled People in Western Society." In *Disability and Society: Emerging Issues and Insights*, edited by Len Barton, 43–52. New York: Longman Publishing, 1996.

Barnes, Colin, and Geof Mercer. "Disability, Work, and Welfare: Challenging the Social Exclusion of Disabled People." *Work, Employment and Society* 19, no. 3 (2005): 527–41.

Barton, Len. “Sociology and Disability: Some Emerging Issues.” In *Disability and Society: Emerging Issues and Insights,* edited by Len Barton, 3–17. New York: Longman Publishing, 1996.

Baynton, Douglas. 2001. “Disability and the Justification of Inequality in American History.” In *The New Disability History,* edited by Paul Longmore and Lauri Umansky, 33–57. New York: New York University Press, 2002.

Bender, Thomas. “The Thinning of American Political Culture.” In *Public Discourse in America,* edited by Judith Rodin and Stephen Steinberg, 27–34. Philadelphia: University of Pennsylvania Press, 2003.

Bengio, Luna. *L'Article 19(1)A de la Loi Canadienne de l'Immigration.* Quebec: Association multi-ethnique pour l'intégration des personnes handicapées du Québec, 1990.

Bensaïd, Daniel. “Permanent Scandal.” In *Democracy In What State?,* edited by Giorgio Agamben et al., 16–43. New York: Columbia University Press, 2011.

Bernhard, Judith K., Patricia Landolt, and Luin Goldring. “Transnationalizing Families: Canadian Immigration Policy and the Spatial Fragmentation of Care-giving among Latin American Newcomers.” *International Migration* 47, no. 2 (2009): 3–30.

Bourne, Larry, and Damaris Rose. 2001. “The Changing Face of Canada: The Uneven Geographies of Population and Social Change.” *The Canadian Geographer* 45, no. 1 (Spring 2001): 105–19.

Boyce, William, Mary Tremblay, Mary Anne McColl, Jerome Bickenbach, Anne Crichton, Steven Andrews, Nancy Gerein, and April D'Aubin. *A Seat at the Table: Persons with Disabilities and Policy Making.* Montreal: McGill-Queen's University Press, 2002.

Boyd, Neil. “Law and Social Control, Law as Social Control.” In *The Social Dimensions of the Law,* edited by Neil Boyd, 2–21. Scarborough, ON: Prentice-Hall Canada, 1986.

Bröckling, Ulrich. “Human Economy, Human Capital: A Critique of Biopolitical Economy.” In *Governmentality: Current Issues and Future Challenges,* edited by Ulrich Bröckling, Susanne Krasmann, and Thomas Lemke, 247–68. New York: Routledge, 2011.

Brown, Wendy. “We Are All Democrats Now...,” in *Democracy In What State?,* edited by Giorgio Agamben et al., 44–57. New York: Columbia University Press, 2011.

Browne, Susan E., Debra Connors, and Nanci Stern. *With the Power of Each Breath: A Disabled Women's Anthology.* Pittsburgh: Cleis Press, 1985.

Bryan, Willie V. *Multicultural Aspects of Disabilities.* Springfield, IL: Charles C. Thomas Publisher, 1999.

Burke, Martin J. *The Conundrum of Class: Public Discourse on the Social Order in America.* Chicago: The University of Chicago Press, 1995.

Byrom, Brad. “A Pupil and a Patient: Hospital-Schools in Progressive America.” In *The New Disability History,* edited by Paul K. Longmore and Lauri Umansky, 133–56. New York: New York University Press, 2001.

Capurri, Valentina. "Women as Individuals and Members of Minority Groups: How to Reconcile Human Rights and the Values of Cultural Pluralism." *GeoJournal* 65, no. 4 (2006): 329–37.

Carter Park, Deborah, and John Radford. "Rhetoric and Place in the 'Mental Deficiency' Asylum." In *Mind and Body Spaces,* edited by Ruth Butler and Hester Parr, 69–96. London: Routledge, 1999.

Chadha, Ena. 2008. "'Mentally Defectives' Not Welcome: Mental Disability in Canadian Immigration Law, 1859–1927." *Disability Studies Quarterly* 28, no.1 (2008).

Chouinard, Vera. "Body Politics: Disabled Women's Activism in Canada and Beyond." In *Mind and Body Spaces,* edited by Ruth Butler and Hester Parr, 267–92. London: Routledge, 1999.

Conrad, Peter and Joseph W. Schneider. *Deviance and Medicalization.* St Louis, MO: The C.V. Mosby Company, 1980.

Council of Canadians with Disabilities. "Make EI Accessible and Inclusive to Canadian Women with Disabilities." Posted 31 March 2009. http://www.ccdonline.ca/en/socialpolicy/employment/EI-pressrelease-31March2009.

Crow, Liz. "Including All of Our Lives: Renewing the Social Model of Disability." *Exploring the Divide: Illness and Disability,* edited by Colin Barnes and Geof Mercer, 55–72. Leeds: The Disability Press, 1996.

Dauvergne, Catherine. "Sovereignty, Migration and the Rule of Law in Global Times." *Modern Law Review* 67, no. 4 (July 2004): 588–615.

Derrida, Jacques. *On Cosmopolitanism and Forgiveness,* London: Routledge, 2001.

Dossa, Parin. *Racialized Bodies, Disabling Worlds: Storied Lives of Immigrant Muslim Women.* Toronto: University of Toronto Press, 2009.

Dowbiggin, Ian Robert. *Keeping America Sane.* Ithaca, NY: Cornell University Press, 2003.

Engler, Yves. *Canada and Israel: Building Apartheid.* Vancouver: RED Publishing, 2010.

Fairchild, Amy L. *Science at the Border,* Baltimore, MD: The John Hopkins University Press, 2003.

Fassin, Didier. "Coming Back to Life: An Anthropological Reassessment of Biopolitics and Governmentality." In *Governmentality. Current Issues and Future Challenges,* edited by Ulrich Bröckling, Susanne Krasmann, and Thomas Lemke, 185–200. New York: Routledge, 2011.

Ferri, Beth A. and David J. Connor. *Reading Resistance.* New York: Peter Lang Publishing, 2006.

Finkel, Alvin. *Social Policy and Practice in Canada: A History.* Waterloo, ON: Wilfrid Laurier University Press, 2006.

Foucault, Michel. *Madness and Civilization.* New York: Vintage Book, 1988. First published 1965 by Vintage (New York).

– *Psychiatric Power: Lectures at the College de France, 1973–1974.* New York: Palgrave MacMillan, 2006.

– *"Society Must Be Defended": Lectures at the College de France, 1975–1976.* New York: Picador, 2003.

Frazee, Catherine. "Exile from the China Shop: Cultural Injunction and Disability Policy." In *Disability and Social Policy in Canada*, edited by Mary Ann McColl and Lyn Jongbloed, 357–69. Concord, ON: Captus Press, 2006.

Freedman, Des. *The Politics of Media Policy.* Cambridge: Policy Press, 2008.

Galer, Dustin. "A Friend in Need or a Business Indeed?: Disabled Bodies and Fraternalism in Victorian Ontario." *Labour/Le Travail* 66 (Fall 2010): 9–36.

– "Packing Disability into the Historian's Toolbox: On the Merits of Labour Histories of Disability." *Labour/Le Travail* 77 (Spring 2016): 256–64.

Gibson, Dale. *The Law of the Charter: Equality Rights.* Toronto: Carswell, 1990.

Ginzberg, Effie. *Power without Responsibility: The Press We Don't Deserve.* Toronto: Urban Alliance on Race Relations, 1987.

Gleeson, Brendan. *Geographies of Disability.* London: Routledge, 1999.

Government of Canada. "Operational Bulletins 037 – Assessing Excessive Demand on Social Services for Business Class Applicants." Last modified 7 September 2007. https://www.canada.ca/en/immigration-refugees-citizenship/corporate/publications-manuals/operational-bulletins-manuals/bulletins-2007/037-september-7-2007.html.

Government of Canada. "Medical Inadmissibility." Last modified 21 December 2018. https://www.canada.ca/en/immigration-refugees-citizenship/services/immigrate-canada/inadmissibility/reasons/medical-inadmissibility.html.

Grekul, Jana, Harvey Krahn, and Dave Odynak. "Sterilizing the 'Feeble-minded': Eugenics in Alberta, Canada, 1929–1972." *Journal of Historical Sociology* 17, no. 4 (December 2004): 358–84.

Grimley Mason, Mary. *Working against Odds: Stories of Disabled Women's Work Lives.* Boston: Northeastern University Press, 2004.

Hackett, Robert A. *News and Dissent: The Press and the Politics of Peace in Canada.* Norwood, NJ: Ablex Pub. Corp, 1991.

Hamilton, Elizabeth C. "From Social Welfare to Civil Rights: The Representation of Disability in Twentieth-Century German Literature." In *The Body and Physical Difference*, edited by David T. Mitchell and Sharon Snyder, 223–39. Michigan: The University of Michigan Press, 1997.

Hawkins, Freda. *Canada and Immigration: Public Policy and Public Concern.* Montreal: McGill-Queen's University Press, 1972.

– *Critical Years in Immigration: Canada and Australia Compared.* Montreal: McGill-Queen's University Press, 1989.

Hayes, David. *Power and Influence.* Toronto: Key Porter Books, 1992.

Henry, Frances. *Discourses of Domination: Racial Bias in the Canadian English-Language Press.* Toronto: University of Toronto Press, 2002.

Holland, Catherine A. *The Body Politic.* New York: Routledge, 2001.

Hughes, Bill, and Kevin Paterson. "The Social Model of Disability and the Disappearing Body: Towards a Sociology of Impairment." *Disability & Society* 12, no. 3 (June 1997): 325–40.

Hunt, Paul. "A Critical Condition." In *The Disability Reader*, edited by Tom Shakespeare, 7–19. London: Continuum, 1998.

Hurst, Rachel. "Conclusion: Enabling or Disabling Globalization?" In *Controversial Issues in a Disabling Society*, edited by John Swain, Sally French, and Colin Cameron, 161–9. Buckingham, PA: Open University Press, 2003.

Hutchinson, Allan C. *Waiting for Coraf: A Critique of Law and Rights*. Toronto: University of Toronto Press, 1995.

Irving, Dan. "Trans Politics and Anti-Capitalism: An Interview with Dan Irving." *Upping the Anti* 4 (2007). https://uppingtheanti.org/journal/article/04-trans-politics-and-anti-capitalism.

Isin, Engin, and Patricia Wood. *Citizenship and Identity*. London: Sage Publications, 1999.

Iyioha, Ireh. "A Different Picture through the Looking-Glass: Equality, Liberalism and the Question of Fairness in Canadian Immigration Health Policy." *Georgetown Immigration Law Journal* 22, no. 4 (Summer 2008): 621–64.

Jayasuriya, Kanishka. *Statecraft, Welfare and the Politics of Inclusion*. Houndmills, UK: Palgrave Macmillan, 2006.

Jennings, Audra. "Engendering and Regendering Disability: Gender and Disability Activism in Postwar America." In *Disability Histories*, edited by Susan Burch and Michael Rembis, 345–63. Urbana: University of Illinois, 2014.

Jongbloed, Lyn. "Disability Policy in Canada: An Overview." *Journal of Disability Policy Studies* 13, no. 4 (January 2003): 203–9.

Kallen, Evelyn. *Ethnicity and Human Rights in Canada*. Toronto: Oxford University Press, 1995.

Karim, Karim H. "Reconstructing the Multicultural Community in Canada: Discursive Strategies of Inclusion and Exclusion." *International Journal of Politics, Culture and Society* 7, no. 2 (December 1993): 189–207.

Kelley, Ninette, and Michael Trebilcock. *The Making of the Mosaic: A History of Canadian Immigration Policy*. Toronto: University of Toronto Press, 1998.

Keshen, Jeff. "Getting It Right the Second Time Around." In *The Veterans Charter and Post-World War II Canada*, edited by Peter Neary and J.L. Granatstein, 62–84. Montreal: McGill-Queen's University Press, 1998.

Larner, Wendy. "Neo-Liberalism: Policy, Ideology, Governmentality." *Studies in Political Economy* 63, no. 1 (Autumn 2000): 5–25.

Lemke, Thomas. "Beyond Foucault: From Biopolitics to the Government of Life." In *Governmentality: Current Issues and Future Challenges*, edited by Ulrich Bröckling, Susanne Krasmann, and Thomas Lemke, 165–84. New York: Routledge, 2011.

Lock, Margaret. "Genomics, Laissez-Faire Eugenics, and Disability." In *Disability in Local and Global Worlds*, edited by Benedicte Ingstad and Susan Reynolds Whyte, 189–211. Berkeley: University of California Press, 2007.

Longmore, Paul K. "Elizabeth Bouvia, Assisted Suicide, and Social Prejudice." In *Why I Burned My Book and Other Essays on Disability*, edited by Paul K. Longmore, 149–74. Philadelphia: Temple University Press, 2003.

Luxemburg, Rosa. "Introduction to Political Economy." In *Economic Writings*, ed. Peter Hudis, 89–300. Vol. 1 of *The Complete Works of Rosa Luxemburg*. London: Verso, 2014.

Malacrida, Claudia. "Discipline and Dehumanization in a Total Institution: Institutional Survivors' Descriptions of Time-Out Rooms." *Disability & Society* 20, no. 5 (2005): 523–37.

– "Income Support Policy in Canada and the UK: Different, but Much the Same." *Disability & Society* 25, no. 6 (2010): 673–86.

Mamann, Guidy. "CIC Skirts Court Medical Ruling." *Metro*, 10 September 2007.

Mamann, Sanaluk & Kingwell LLP. Immigration Lawyers (website). Copyright 1987–2019. https://www.migrationlaw.com.

Manfredi, Christopher P. *Judicial Power and the Charter: Canada and the Paradox of Liberal Constitutionalism*. Don Mills, ON: Oxford University Press, 2001.

Mayr, Andrea. "Introduction: Power, Discourse and Institutions." In *Language and Power*, edited by Andrea Mayr, 53–76. London: Continuum International Publishing Group, 2008.

McLaren, Angus. *Our Own Master Race*. Toronto: McClelland & Stewart, 1990.

Menzies, Robert. "Governing Mentalities: The Deportation of 'Insane' and 'Feebleminded' Immigrants Out of British Columbia from Confederation to World War II." *Canadian Journal of Law and Society* 13, no. 2 (Fall 1998): 135–76.

Miliband, Ralph. *The State in Capitalist Society*. London: Quartet Books, 1973.

Moran, James E. *Committed to the State Asylum: Insanity and Society in Nineteenth-Century Quebec and Ontario*. Montreal: McGill-Queen's University Press, 2000.

Morris, Jenny. *Pride Against Prejudice. Transforming Attitudes to Disability*. London: The Women's Press, 1991.

Myles, John, and Feng Hou. "Changing Colours: Spatial Assimilation and New Racial Minority Immigrants." *Canadian Journal of Sociology* 29, no. 1 (2004): 29–58.

Neufeld, Aldred. "Growth and Evolution of Disability Advocacy in Canada." In *Making Equality*, edited by Deborah Stienstra and Aileen Wight-Felske, 11–32. Toronto: Captus Press, 2003.

– "Disability Policy (Canada)." In *Encyclopedia of Social Welfare History in North America*, edited by John M. Herrick and Paul H. Stuart, 78–80. Thousand Oaks, CA: Sage Publications, 2005.

Oliver, Michael. "Disability and Dependency: A Creation of Industrial Societies?" In *Disability and Dependency*, edited by Len Barton, 7–22. London: The Falmer Press, 1989.

– *The Politics of Disablement.* Houndmills, UK: Macmillan, 1990.
– *Understanding Disability.* New York: St Martin's Press, 1996.
Pâquet, Martin. *Tracer les marges de la cite: Étranger, immigrant et l'État au Québec, 1627–1981.* Montreal: Boréal, 2005.
Peters, Yvonne. "From Charity to Equality." In *Making Equality,* edited by Deborah Stienstra and Aileen Wight-Felske, 119–36. Toronto: Captus Press, 2003.
Philo, Greg. "Can Discourse Analysis Successfully Explain the Content of Media and Journalistic Practice?" *Journalism Studies* 8, no. 2 (April 2007): 175–96.
Plant, Raymond. *The Neo-Liberal State.* Oxford: Oxford University Press, 2010.
Prince, Michael J. "Claiming a Disability Benefit as Contesting Social Citizenship." In *Contesting Illness: Processes and Practices,* edited by Pamela Moss and Katherine Teghtsoonian, 28–46. Toronto: University of Toronto Press, 2008.
– *Absent Citizens: Disability Politics and Policy in Canada.* Toronto: University of Toronto Press, 2009.
Proctor, Robert. *Racial Hygiene: Medicine under the Nazis.* Cambridge, MA: Harvard University Press, 1988.
Qadeer, Mohammad, and Sandeep Kumar. "Ethnic Enclaves and Social Cohesion." In "Our Diverse Cities: Challenges and Opportunities," edited by John Biles, Erin Tolley, and Jim Zamprelli. Special Issue, *Canadian Journal of Urban Research* 15, no. 2 (2006): 1–17.
Razack, Sherene. *Looking White People in the Eye: Gender, Race, and Culture in Courtrooms and Classrooms.* Toronto: University of Toronto Press, 1998.
Reaume, Geoffrey. "Patients at Work: Insane Asylum Inmates' Labour in Ontario, 1841–1900." In *Mental Health and Canadian Society: Historical Perspectives,* edited by James E. Moran and David Wright, 69–99. Montreal: McGill-Queen's University Press, 2006.
Roberts, David J., and Minelle Mahtani. "Neoliberalizing Race, Racing Neoliberalism: Placing 'Race' in Neoliberal Discourses." *Antipode* 42, no. 2 (March 2010): 248–57.
Rodgers, Daniel T. *The Work Ethic in Industrial America 1850–1920,* Chicago: The University of Chicago Press, 1978.
Rooke, Patricia T., and R.L. Schnel. *Discarding the Asylum: From Child Rescue to the Welfare State in English-Canada (1800–1950).* Lanham, MD: University Press of America, 1983.
Rose, Martha L. *The Staff of Oedipus.* Ann Arbor: The University of Michigan Press, 2003.
Rosen, Jay. "Part of Our World: Journalism as Civic Leadership." In *Public Discourse in America,* edited by Judith Rodin and Stephen P. Steinberg, 106–17. Philadelphia: University of Pennsylvania Press, 2003.
Roy, Patricia E. *The Triumph of Citizenship.* Vancouver: UBC Press, 2007.
Rutherford, Paul. *The Making of the Canadian Media.* Toronto: McGraw-Hill Ryerson Limited, 1978.

Ryfe, David M. "The Principles of Public Discourse: What Is Good Public Discourse?" In *Public Discourse in America,* edited by Judith Rodin and Stephen P. Steinberg, 163–77. Philadelphia: University of Pennsylvania Press, 2003.

Said, Edward W. *Covering Islam: How the Media and the Experts Determine How We See the Rest of the World.* New York: Pantheon Books, 1981.

Sandys, Judith. "Immigration and Settlement Issues for Ethno-Racial People with Disabilities: An Exploratory Study." Updated 1 September 1998. http://ceris.ca/wp-content/uploads/virtual-library/Sandys_1999.pdf.

Sayad, Abdelmalek. *La double absence,* Paris: Seuil, 1999.

Sears, Alan. "Immigration Controls as Social Policy: The Case of Canadian Medical Inspection 1900–1920." *Studies in Political Economy* 33 (1990): 91–112.

Shearer, Ann. *Disability: Whose Handicap?* Oxford: Basil Blackwell Publisher, 1981.

Simmons, Alan. "Immigration Policy: Imagined Futures." In *Immigrant Canada: Demographic, Economic, and Social Challenges,* edited by Shiva S. Halli and Leo Driedger, 21–50. Toronto: University of Toronto Press, 1999.

Smandych, Russell C., and Simon N. Verdun-Jones. "The Emergence of the Asylum in 19th Century Ontario: A Study in the History of Segregative Control." In *The Social Dimensions of Law,* edited by Neil Boyd, 166–81. Scarborough, ON: Prentice-Hall Canada, 1986.

Smart, Barry. *Economy, Culture and Society,* Buckingham, PA: Open University Press, 2003.

Smelser, Neil. "A Paradox of Public Discourse and Political Democracy." In *Public Discourse in America,* edited by Judith Rodin and Stephen P. Steinberg, 178–83. Philadelphia: University of Pennsylvania Press, 2003.

Smith, William George. *Building the Nation: A Study of Some Problems Concerning the Churches' Relation to the Immigrants.* Toronto: Canadian Council of the Missionary Education Movement, 1922.

Statistics Canada. *Canada's Ethnocultural Mosaic, 2006 Census.* Ottawa: Minister of Industry, 2008.

Stone, Deborah A. *The Disabled State.* Philadelphia: Temple University Press, 1984.

Stone, Sharon Dale. "Resisting an Illness Label: Disability, Impairment, and Illness." In *Contesting Illness: Processes and Practices,* edited by Pamela Moss and Katherine Teghtsoonian, 201–17. Toronto: University of Toronto Press, 2008.

Suzuki, David. *The David Suzuki Reader.* Vancouver: Greystone Books, 2003.

Swain, John, Sally French, and Colin Cameron. *Controversial Issues in a Disabling Society.* Buckingham, PA: Open University Press, 2003.

Thobani, Sunera. *Exalted Subjects: Studies in the Making of Race and Nation in Canada.* Toronto: University of Toronto Press, 2007.

Thomson, Rosemarie Garland. *Extraordinary Bodies: Figuring Physical Disability in American Culture and Literature.* New York: Columbia University Press, 1997.

Timlin, Mabel F. "Canada's Immigration Policy, 1896–1910." *The Canadian Journal of Economics and Political Science* 26, no. 4 (1960): 517–32.
Titchkosky, Tanya. *Disability, Self, and Society*, Toronto: University of Toronto Press, 2003.
– *Reading and Writing Disability Differently: The Textured Life of Embodiment.* Toronto: University of Toronto Press, 2007.
Tregaskis, Claire. "Social Model Theory: The Story so Far ..." *Disability & Society* 17, no. 4 (2002): 457–70.
Trent, James W. Jr. *Inventing the Feeble Mind.* Los Angeles: University of California Press, 1994.
Troper, Harold. *History of Immigration since the Second World War: From Toronto "The Good" to Toronto "The World in a City."* Toronto: Joint Centre of Excellence for Research on Immigration and Settlement, 2000.
Turner, Bryan S. *The Body and Society.* London: Sage Publications, 1996.
United Nations. "Standard Rules on the Equalization of Opportunities for Persons with Disabilities." Enable. Copyright 2006. http://www.un.org/esa/socdev/enable/dissre01.htm#Background.
van Dijk, Teun A. "Principles of Critical Discourse Analysis." *Discourse & Society* 4, no. 2 (April 1993): 249–83.
Voyvodic, Rose. "Into the Wasteland: Applying Equality Principles to Medical Inadmissibility in Canadian Immigration Law." *Journal of Law and Social Policy* 16 (2001): 115–44.
Waddell, Bill. "United States Immigration: A Historical Perspective." In *Handbook of Immigrant Health*, edited by Sana Loue, 1–17. New York: Plenum Press, 1998.
Walks, R. Alan, and Larry S. Bourne. "Ghettos in Canadian Cities? Racial Segregation, Ethnic Enclaves and Poverty Concentration in Canadian Urban Areas." *The Canadian Geographer* 50, no. 3 (2006): 273–97.
Wendell, Susan. *The Rejected Body.* New York: Routledge, 1996.
Whitaker, Reg. *Canadian Immigration Policy since Confederation.* Ottawa: Canadian Historical Association, 1991.
Wolfensberger, Wolf. *The Origin and Nature of Our Institutional Models.* New York: Human Policy Press, 1975.
Wood, Patricia. "Defining "Canadian": Anti-Americanism and Identity in Sir John Macdonald's Nationalism." *Journal of Canadian Studies* 36, no. 2 (2001): 49–69.
Woodsworth, James Shaver. *Strangers within Our Gates.* Toronto: University of Toronto Press, 1972. First published 1908 by F.C. Stephenson (Toronto).
Yingling, Thomas E. *AIDS and the National Body.* Durham, NC: Duke University Press, 1997.
Young, Iris M. *Justice and the Politics of Difference.* Princeton, NJ: Princeton University Press, 1990.

Zinn, Howard. *Failure to Quit: Reflections of an Optimistic Historian.* Cambridge: South End Press, 2002.

– *Howard Zinn Speaks: Collected Speeches 1963–2009.* Chicago: Haymarket Books, 2012.

Legislation, Legal Cases, and Government Documents

Canadian Charter of Rights and Freedoms. https://laws-lois.justice.gc.ca/eng/const/page-15.html

Canada Constitution Act 1982. Last modified 6 April 2019. http://lawslois.justice.gc.ca/eng/const/.

Canada. House of Commons Debates. Ottawa: Queen's Printer for Canada. 1867–.

Canada. Minutes of Proceedings and Evidence of the Legislative Committee on Bill C-86, An Act to Amend the Immigration Act and Other Acts in Consequence thereof. 1992. Ottawa: Queen's Printer for Canada.

Canada. Senate Debates. Ottawa: Queen's Printer for Canada. 1869–2005.

Canada. Sessional Papers of the Parliament of Dominion of Canada. Ottawa: Hunter, Rose. 1869–2005.

Canada. Standing Committee on Citizenship and Immigration. 8 June 2000. http://www.ourcommons.ca/DocumentViewer/en/36-2/CIMM/meeting-30/evidence.

Canada. Standing Committee on Citizenship and Immigration. 29 March 2000. http://www.ourcommons.ca/DocumentViewer/en/36-2/CIMM/meeting-19/evidence.

Canada. Standing Committee on Citizenship and Immigration. 1 May 2001. http://www.ourcommons.ca/DocumentViewer/en/37-1/CIMM/meeting-14/evidence.

Canada. Standing Committee on Citizenship and Immigration. 5 February 2002. http://www.ourcommons.ca/DocumentViewer/en/37-1/CIMM/meeting-45/evidence.

Canada. Standing Committee on Citizenship and Immigration. 12 March 2002. http://www.ourcommons.ca/DocumentViewer/en/37-1/CIMM/meeting-53/evidence.

Chesters v. Canada (Minister of Citizenship and Immigration), No. IMM-1316-97. Federal Court Trial Division, 27 June 2002.

Chesters v. Canada. Written Submissions of the Intervener, Council of Canadians with Disabilities, No. IMM-1316-97. Federal Court Trial Division, 27 February 2001.

De Jong v. Canada (Minister of Citizenship and Immigration), Docket IMM-6058-99. Federal Court Trial Division, 13 November 2002.

De Jong v. Canada (Minister of Citizenship and Immigration), No. 30127. Supreme Court of Canada, 21 October 2005.

Deol v. Canada (Minister of Employment and Immigration), No. A-280-90. Federal Court of Appeal, 27 November 1992.
Gingiovenanu v. Canada (Minister of Employment and Immigration), No. IMM-3875-93. Federal Court Trial Division, 30 October 1995.
Hilewitz v. Canada (Minister of Citizenship and Immigration), No. IMM-5340-00. Federal Court Trail Division, 8 August 2002.
Hilewitz v. Canada (Minister of Citizenship and Immigration), No. A-560-02. Federal Court of Appeal, 12 November 2003.
Hilewitz v. Canada (Minister of Citizenship and Immigration), No. 30125. Supreme Court of Canada, 21 October 2005.
Immigration and Refugee Protection Act. Last modified 26 June 2019. https://laws.justice.gc.ca/eng/acts/i-2.5/.
Ismaili v. Canada (Minister of Citizenship and Immigration), No. IMM-3430-94. Federal Court Trial Division, 17 August 1995.
Lau v. Canada (Minister of Citizenship and Immigration), No. IMM-4361-96. Federal Court Trial Division, 17 April 1998.
Pamar v. Canada (Minister of Employment and Immigration), No. A-836-37. Federal Court of Appeal, 16 May 1988.
Poon v. Canada (Minister of Citizenship and Immigration), No. IMM-2007-99. Federal Court Trial Division, 1 December 2000.
Poste v. Canada (Minister of Citizenship and Immigration), No. IMM-4601-96. Federal Court Trial Division, 22 December 1997.
Redding v. Canada (Minister of Citizenship and Immigration), No. IMM-2661-00. Federal Court Trial Division, 22 August 2001.
Wong v. Canada (Minister of Citizenship and Immigration), No. IMM-6060-99. Federal Court Trial Division, 31 May 2002.

Index